Series FM1 no.32

Birth Statistics

Review of the Registrar General on births and patterns of family building in England and Wales, 2003

ISBN 1 85774 593 0
ISSN 0140–2587

Contact points
For enquiries about this publication, contact
Vital statistics Ouputs Branch
Tel: 01329 813758
E-mail: vsob@ons.gov.uk

For general enquiries, contact the National Statistics Customer Contact Centre on: 0845 601 3034
(minicom: 01633 812399)
E-mail: info@statistics.gsi.gov.uk
Fax: 01633 652747
Post: Room 1015, Government Buildings,
Cardiff Road, Newport NP10 8XG

You can also find National Statistics on the Internet at:
www.statistics.gov.uk

About the Office for National Statistics
The Office for National Statistics (ONS) is the government agency responsible for compiling, analysing and disseminating many of the United Kingdom's economic, social and demographic statistics, including the retail prices index, trade figures and labour market data, as well as the periodic census of the population and health statistics. It is also the agency that administers the statutory registration of births, marriages and deaths in England and Wales. The Director of ONS is also the National Statistician and the Registrar General for England and Wales.

A National Statistics publication
National Statistics are produced to high professional standards set out in the National Statistics Code of Practice. They undergo regular quality assurance reviews to ensure that they meet customer needs. They are produced free from any political influence.

Contents

List of main tables and appendices

5 Marriage duration

6 Multiple births

7 Area of usual residence

1 Introduction

Birth statistics 2003 presents statistics on births occurring annually in England and Wales between 1993 and 2003.

The 2003 annual commentary on fertility trends will be published in *Population Trends 118* in December 2004. Provisional data for the complete year are first published as Reports in *Population Trends* as soon as they become available.

This volume is produced by the Office for National Statistics (ONS). It is published under the National Statistics logo, the designation guaranteeing that those outputs have been produced to high professional standards set out in a Code of Practice, and have been produced free from political interference.

The registration of life events (births, deaths and marriages) is a service carried out by the Local Registration Service in partnership with the General Register Office (GRO) in Southport, which is part of ONS. ONS was formed on 1 April 1996, bringing together the Office of Population Censuses and Surveys (OPCS) and the Central Statistical Office. OPCS is referred to in this volume for historic events and publications.

1.1   Tables in this volume

The tables presented in this volume are set out in 11 sections, covering various fertility topics including characteristics of births and of the parents. **Tables 10.1** to **10.5** analyse trends in cohort fertility; otherwise, the tables use period fertility measures.

For brevity, the time series shown here have been limited to a run of 11 years at most. Figures for earlier years are shown in earlier volumes of *Birth statistics*. This series includes a volume of historical fertility statistics[1] containing some time series back to 1837, the year of the introduction of compulsory birth registration. More detailed time series are also included for years from 1938, the year in which the first Population Statistics Act came into force - see section 2.14.

As described in *Birth Statistics 2002*, conception figures are now published separately in a supplementary volume. The supplementary volume *Conception Statistics 2001* will be published in December 2004.

Table 1.5 shows number of marriages and marriage rates for the years 1992 to 2002. The figures for 2002 are provisional.

Provisional figures for 2002 are also used for mean age of women at marriage in **Table 1.6**.

For **Tables 7.1** to **7.7, 8.2, 8.3** and **9.2,** the presentation of figures for administrative areas has been changed in line with guidance for the presentation of government statistics for regional and sub-regional statistics.

'England and Wales' replaces 'England, Wales and Elsewhere' and includes all births occurring in England and Wales.

'Elsewhere' is re-named 'Normal residence outside England and Wales' and is now placed at the end of each table with the relevant counts.

Total counts for Wales are now not shown separately at the beginning of each table, as they now appear above all the counts for areas in Wales (Local Health Boards in **Tables 7.2, 7.4, 8.2** and **8.3**). These also appear in geographical instead of alphabetical order.

Table 9.5 shows total fertility rates by mother's country of birth and uses population denominators from the Censuses of 1991 and 2001. It aggregates countries into groupings that provide denominators large enough for reliable rates. Surveys conducted between Censuses are unable to provide reliable denominators for intervening years.

Data from 2001 onwards have been coded to the new socio-economic classification as defined by occupation and employment status. Therefore **Tables 11.1** to **11.5** have been adjusted to show data for the three years 2001, 2002 and 2003 only. See section 3.10 for further information.

1.2   Data analysed in this volume

The information used in tables in this volume is based largely on the details collected when births are registered. Most of the information, for both live births and stillbirths, is supplied to registrars by one or both parents. For live births, details of **birthweight** are notified to the local health authority by the hospital where the birth took place, or by the midwife or doctor in attendance at the birth. These details are then supplied to the registrar. For stillbirths, details of **cause of death, duration of pregnancy and weight of foetus** are supplied on a certificate or notification by the doctor or midwife either present at the birth, or who examined the body. The certificate or notification is then taken by the informant to a registrar. The registrar will use all this information to complete a draft entry *Form 309* (**Annex A**) for a live birth or *Form 308* (**Annex B**) for a stillbirth.

The data used in this volume are summarised below. The items covered are collected for both live births and stillbirths.

Table number	Year(s)	Area	Numbers/ rates/ percent- ages (N/R/P)	Live births/ stillbirths (LB/SB)	Mater- nities/ pater- nities (M/P)	Age of mother (Year of birth)	Age of father	Period of occur- rence	Period of regist- ration	Sole/ joint registra- tion (S/J)	Within/ outside marriage (W/O)
	Summary										
1.1	1993-2003	E&W	NR	LB							WO
1.2	1993-2003	E&W	NR	SB							WO
1.3	1993-2003	E&W	NR	LB							
1.4	1993-2003	E&W	R	LB							
1.5	1992-2002	E&W	NR								
1.6	1993-2003	E&W	Mean	LB		*					WO
1.7	1993-2003	E&W	Mean	LB		*					
1.8	1992-2002	E&W	P	LB/SB		*					W
1.9	1993-2003	E&W	R	LB		*					WO
	Seasonality										
2.1	1993-2003	E&W	NR	LB					*		
2.2	1993-2003	E&W	N	SB					*		
2.3	1993-2003	E&W	NR	LB					*		
2.4	2003	E&W	N	LB/SB	M				*		WO
2.5	2003	E&W	N	LB					*	*	
	Parents' age										
3.1	1993-2003	E&W	NR	LB		*					WO
3.2	2003	E&W	N	LB/SB	M	*					WO
3.3	1993-2003	E&W	NR	LB			*				W
3.4	2003	E&W	N	LB/SB	P		*			J	WO
3.5	2003	E&W	R	LB/SB	P		*				W
3.6	2003	E&W	N	LB		*	*				W
3.7	2003	E&W	N	SB		*	*				W
3.8	2003	E&W	N	LB		*	*			SJ	O
3.9	1993-2003	E&W	NP	LB		*				SJ	O
3.10	1993-2003	E&W	NP	LB		*				J	O
	Previous children										
4.1	1993-2003	E&W	N	LB		*					W
4.2	2003	E&W	N	LB		*					W
4.3	2003	E&W	NR	LB/SB		*					W
	Marriage duration										
5.1	1993-2003	E&W	N	LB		*					W
5.2	1993-2003	E&W	N	LB		*					W
5.3	2003	E&W	N	LB		*					W
	Multiple births										
6.1	1993-2003	E&W	NR		M	*					WO
6.2	2003	E&W	N	LB/SB	M	*					
6.3	2003	E&W	NR		M	*					W
6.4	2003	E&W	N	LB/SB	M	*					
	Area										
7.1	2003	E&W+	NR	LB							WO
7.2	2003	E&W+	NR	LB							WO
7.3	2003	E&W+	NR	LB		*					
7.4	2003	E&W+	NR	LB		*					
7.5	2003	E&W+	NP	LB							
7.6	2003	E&W+	NR	SB							WO
7.7	2003	E&W+	NP	SB							

Table number	Year(s)	Area	Numbers/ rates/ percent- ages (N/R/P)	Live births/ stillbirths (LB/SB)	Mater- nities/ pater- nities (M/P)	Age of mother (Year of birth)	Age of father	Period of occur- rence	Period of regist- ration	Sole/ joint registra- tion (S/J)	Within/ outside marriage (W/O)
	Place of confinement										
8.1	2003	E&W	N		M	*					WO
8.2	2003	E&W+	N		M						
8.3	2003	E&W+	NR	LB/SB	M						
	Parents' birthplace										
9.1	1993, 1998-2003	E&W	NP	LB							
9.2	2003	E&W+	NP	LB							
9.3	2003	E&W	N	LB							
9.4	2003	E&W	N	LB		*					
9.5	1991 and 2001	E&W	R	LB							
9.6	1993, 2001-2003	E&W	NP	LB							WO
	Cohorts										
10.1	1920-1988	E&W	R	LB		*					
10.2	1920-1988	E&W	R	LB		*					
10.3	1920-1988	E&W	R	LB		*					
10.4	1920-1984	E&W	R	LB		*		*			WO
10.5	1920-1983	E&W	P	LB		*					
	Socio-economic classification										
11.1	2001-2003	E&W	N	LB		*					W
11.2	2001-2003	E&W	NP	LB		*					W
11.3	2001-2003	UK and E&W	Median	LB							W
11.4	2001-2003	E&W	Mean	LB		*		*			W
11.5	2001-2003	E&W	N	LB		*				J	O

The **date of birth** is supplied in a conventional way, except that where more than one child is liveborn at a confinement, then time is also recorded. **Place of birth** is entered as the usual name and the address of a hospital, maternity home or other communal establishment, or the address of a private dwelling. ONS then codes place of birth to one of the groups of places in **Table 8.1.** The **sex** of the child is also recorded.

Although the **birthplace of the parents** may be recorded in detail if this was in the United Kingdom, the main interest in this volume is with parents born outside the United Kingdom - see **Tables 9.1** to **9.6.** The **mother's usual address** is entered, as is that of the informant where appropriate. This information is used for tables showing usual residence of mother **7.1** to **7.7**, as well as **Table 3.10,** which analyses jointly registered births outside marriage and whether the parents resided at the same address.

Occupation is recorded for the mother and for the father, if his name is entered in the register. The informant is asked whether the father/mother was in gainful employment at any time before the child's birth, and a description of the occupation may be recorded. The informant may not wish to have details of the father/mother's occupation entered in the register, but it may still be recorded for use in statistical analyses. If the father is unemployed, his last full-time occupation will be recorded. As discussed in section 3.10, this information is used for analyses of socio-economic classification as defined by occupation in **Tables 11.1** to **11.5.**

Informants are also required to provide further information, treated as confidential, under the provisions of the Population Statistics Acts, as below:

(i)   the **father's date of birth,** if his name is entered in the register when the registration is joint;
(ii) the **mother's date of birth;**

if the child's parents were married to each other at the time of the birth:

(iii) the **date of the parents' marriage;**
(iv) **whether the mother has been married more than once;**
(v)   **number of previous children** by her present husband and any former husband, (a) **born alive,** and (b) **stillborn.**

These confidential details are used extensively in this volume, in particular for analyses of **age of mother and father,** of **marriage duration,** and of **birth order.** See section 1.3 below for details of issues affecting the quality of these variables in 2003.

Sex	Previous live-born children	Birth-weight	Place of confine-ment	Country of birth mother/ father (M/F)	Socio-economic classification of father/ husband	Usual residence/ place of occurrence (R/O)	Marriage order	Year(s)	Table number
								Place of confinement	
	*		*					2003	8.1
			*			RO		2003	8.2
						O		2003	8.3
								Parents' birthplace	
				M				1993, 1998-2003	9.1
				M		R		2003	9.2
				MF				2003	9.3
				M				2003	9.4
				M				1991 and 2001	9.5
	*			M				1993, 2001-2003	9.6
								Cohorts	
								1920-1988	10.1
								1920-1988	10.2
								1920-1988	10.3
	*							1920-1984	10.4
	*							1920-1983	10.5
								Socio-economic classification	
	*				*			2001-2003	11.1
					*			2001-2003	11.2
					*			2001-2003	11.3
	*				*			2001-2003	11.4
					*			2001-2003	11.5

Other statistical information collected at registration includes the economic activity of the parents, i.e. **industry** and **employment status,** and whether the confinement resulted in a **multiple birth.**

1.3   Issues affecting the quality of the data in this volume

After the publication of the 1997 volume in this series, an error was discovered in the births database that resulted in 1,002 live births being excluded from the published 1997 statistics. Details of the error were included in section D.1 of the 1998 volume[2]. All time series tables in this and other volumes since 1997 reflect the corrected data for 1997.

As noted above, some statistical information is collected under the Population Statistics Acts at the registration of a birth. In 1999 the proportion of live birth registrations without this information received from one Register Office was higher than usual due to a combination of circumstances. The missing data on those records were imputed using a random sample of data from the particular area from the previous three years. This was a change from the usual method, but was used to improve the quality of the imputations. Procedures were put in place which mean that such a problem is unlikely to recur. For further information about this see section C.2 of the 1999 volume[3].

Since the dataset for the 2000 volume, a small number of very late registrations has been excluded each year from the official statistics. Inclusion of these very late registrations in the statistical dataset was found to have an adverse effect on the quality of infant mortality data when linked with the live birth data. The annual dataset now includes only those births occurring in the reference year, and late registrations of births occurring in the year previous to the reference year. See section 2.2 for further discussion of occurrences and registrations. The numbers of late registrations included in, and numbers of very late registrations excluded from the statistics are shown in the table below:

Annual dataset year	Number of late registrations from the previous year included in the dataset	Number of very late registrations excluded from the dataset
2000	519	34
2001	195	20
2002	161	17
2003	207	15

1.4 Associated publications and the National Statistics website

The National Statistics website (**www.statistics.gov.uk**) provides a comprehensive source of freely available vital statistics and ONS products. More information on the National Statistics website can be obtained from the contact address in section 2.16.

Historic data, and figures for UK countries

Comparable statistics for earlier years, and separate statistics for Scotland and Northern Ireland are published as follows:

England and Wales: from 1974-2001 in *Birth statistics;* for earlier years in the *Registrar General's Statistical Review of England and Wales.* Data for the years 1837-1983 are summarised in an earlier volume in the FM1 series[1].

Scotland: in the *Annual Report of the Registrar General for Scotland.*

Northern Ireland: in the *Annual Report of the Registrar General for Northern Ireland.*

A summary of fertility statistics for the United Kingdom and constituent countries appears in the *Annual Abstract of Statistics,* issued by ONS. Similar data also appear in the ONS quarterly journals *Population Trends* and *Health Statistics Quarterly.* Data for Europe are published in the Council of Europe annual volume *Recent demographic developments in Europe,* and the Eurostat publication *Population Statistics.* Statistics for United Nations member countries appear in the annual *UN Demographic Yearbook.*

Other related annual reference volumes published by ONS include:

- **DH3** *(Mortality statistics: childhood, infant and perinatal),* which contains data on stillbirths, infant deaths, and childhood deaths. It includes figures for infant deaths linked to their corresponding birth records, as well as birthweight data for health authorities analysed separately by age, socio-economic classification and parity.

- **FM2** *(Marriage, divorce and adoption statistics),* which provides data on marriages by age of bride and bridegroom, on divorces by ages of children involved, and on adoptions by age of child at adoption.

Fertility data are also published annually in the ONS volume *Key population and vital statistics*[4]. This volume provides data on population, births, deaths, and migration, for administrative areas in the United Kingdom, including local and health authorities.

Population Trends and Health Statistics Quarterly publications

Up to 1998 ONS published annual data in Monitors, known as the FM1 Series for conceptions and live births. These contained basic information on annual conceptions and live birth registrations, and were issued soon after the data became available. However, these publications have been discontinued and since 1999 these data have appeared in Reports issued in the quarterly journal *Population Trends.* Since the beginning of 1999, ONS has published two quarterly journals: *Population Trends,* which now has an emphasis on population and demography, covering most fertility topics, and *Health Statistics Quarterly,* covering mortality and health topics, including abortions, and some other fertility data. The annual Report on live births by local and health authority areas is published in the June issue of *Population Trends.*

Population Trends and *Health Statistics Quarterly* both contain regular quarterly reference tables on a variety of population and health topics; for fertility these include analyses of births (numbers and rates) by age of mother.

1.5 Other publications

Some other recent background information on fertility data and other relevant articles and publications are listed below. Most are from the journal *Population Trends,* but copies of any not easily available may be obtained from ONS - see section 2.16.

Trends in fertility

- ONS. Annual update: Births in 2003 in England and Wales, *Population Trends* 118, pp 72-76, Winter 2004.
- Berrington, A. Perpetual postponers? Women's, men's and couple's fertility intentions and subsequent fertility behaviour. *Population Trends* 117, pp 9-19, Autumn 2004.
- Smallwood, S. Characteristics of sole registered births and the mothers who register them. *Population Trends* 117, pp 20-26, Autumn 2004.
- ONS. Annual update: Births in 2002 in England and Wales. *Population Trends* 115, pp 83-86, Spring 2004.
- Kiernan, K and Smith, K. Unmarried parenthood: new insights from the Millennium Cohort Study. *Population Trends* 114, pp 26-33, Winter 2003.
- Murphy, M and Grundy, E. Mothers with living children and children with living mothers: the role of fertility and mortality in the period 1911-2050. *Population Trends* 112, pp 36-44, Summer 2003.
- ONS. Annual update: Births in 2001 and Conceptions in 2000 in England and Wales. *Population Trends* 110, pp 78-81, Winter 2002.

- Smallwood, S. The effects of changes in timing of childbearing on measuring fertility in England and Wales. *Population Trends* 109, pp 36-45, Autumn 2002.
- Joshi, H and Smith, K. The millennium cohort study. *Population Trends* 107, pp 30-34, Spring 2002.
- Botting, B and Dunnell, K. Trends in fertility and contraception in the last quarter of the 20th century. *Population Trends* 100, pp 32-40, Summer 2000.
- Armitage, R and Babb, P. Population Review (4): Trends in fertility. *Population Trends* 84, pp 7-13, Summer 1996.

Age patterns

- Rendall, M. How important are inter-generational cycles of teenage motherhood in England and Wales? A comparison with France. *Population Trends* 111, pp. 27-37, Spring 2003.
- Babb, P. Fertility of the over forties. *Population Trends* 79, pp 34-36, Spring 1995.
- Babb, P. Teenage conceptions and fertility in England and Wales, 1971-91. *Population Trends* 74, pp 12-17, Winter 1993

Socio-economic classification

- Donkin, A, Lee Y and Toson, B. Implications of changes in the UK social and occupational classifications in 2001 for vital statistics. *Population Trends* 107, pp 23-29, Spring 2002.

Birth order

- Rendall, M and Smallwood, S. Higher qualifications, first-birth timing and further childbearing in England and Wales. *Population Trends* 111, pp 18-26, Spring 2003.
- Smallwood, S. New estimates of trends in birth order in England and Wales. *Population Trends* 108, pp 32-48, Summer 2002.
- Wood, R. Trends in multiple births 1938-1995. *Population Trends* 87, pp 29-35, Spring 1997.

Future levels of fertility

- Shaw, C. Interim 2003-based national population projections for the United Kingdom and constituent countries. *Population Trends* 118,pp 6-17, Winter 2004.
- Shaw, C. 2002-based national population projections for the United Kingdom and constituent countries. *Population Trends* 115, pp 6-15, Spring 2004.
- Smallwood, S. Fertility assumptions for the 2002-based national population projections. *Population Trends* 114, pp 8-18, Winter 2003.

- Shaw, C. Interim 2001-based national population projections for the United Kingdom and constituent countries. *Population Trends* 111, pp 7-17, Spring 2003
- GAD. National population projections 2000–based (Series PP2 no 23). Fertility, Chapter 6, pp 18-21, The Stationery Office, 2002.
- Shaw, C. 2000-based national population projections for the United Kingdom and its constituent countries. *Population Trends* 107, pp 5-13, Spring 2002.
- Shaw, C. Assumptions for the 2000-based National Population Projections. *Population Trends* 105 pp 45-47, Autumn 2001.
- GAD. 1998-based national population projections (Series PP2 no 22). Fertility, Chapter 7, pp 23-26, The Stationery Office, 2000.
- Shaw, C. 1996-based national population projections for the United Kingdom and constituent countries. *Population Trends* 91, pp 43-49, Spring 1998.
- Craig, J. Replacement level fertility and future population growth. *Population Trends* 78, pp 20-22, Winter 1994.

Other

- Hancock, R, Stuchbury, R and Tomassini, C. Changes in the distribution of marital age differences in England and Wales, 1963 to 1998. *Population Trends* 114, pp 19-25, Winter 2003.
- Smith, J, Chappell, R, Whitworth, A and Duncan, C. Implications of 2001 Census for local authority district mid-year population estimates. *Population Trends* 113, pp 20-31, Autumn 2003.
- Smallwood, S and Jefferies, J. Family building intentions in England and Wales: trends, outcomes and interpretations. *Population Trends* 112, pp 15-28, Summer 2003.
- Hindess, G. Population review of England and Wales, 2001. *Population Trends* 112, pp 7-14, Summer 2003.
- Ghee, C. Population review of 2000, England and Wales. *Population Trends* 106, pp 7-14, Winter 2001.
- Berthoud, R. Teenage births to ethnic minority women. *Population Trends* 104, pp 12-18, Summer 2001.
- Haskey, J. Having a birth outside marriage: the proportion of lone mothers and cohabiting mothers who subsequently marry. *Population Trends* 97, pp 6-18, Autumn 1999.
- Botting, B, Rosato, M and Wood, R. Teenage mothers and the health of their children. *Population Trends* 93, pp 19-28, Autumn 1998.
- Filakti, H. Trends in abortion 1990-1995. *Population Trends* 87, pp 11-19, Spring 1997.
- Armitage, R. Variation in fertility between different types of local area. *Population Trends* 87, pp 20-28, Spring 1997.

2   Notes and definitions

2.1   Base populations

The population figures in **Appendix Table 1**, which are used to calculate fertility rates in this volume, are mid-year estimates of the resident population of England and Wales based on the 2001 Census of Population. These estimates include members of HM and non-UK armed forces stationed in England and Wales, but exclude those stationed outside. ONS mid-year population estimates are updated figures using the most recent Census, allowing for births, deaths, net migration and ageing of the population.

Whenever results become available from a Census, a new base is created for the population estimates. Thus, following the 2001 Census, ONS revised the mid-year population estimates from 1982 to 2000. Further revisions, though, subsequently occurred to the estimates from 1992. This is because further information from research following the Census became available. The research was carried out to understand the reasons for differences between the 2001 Census-based estimates and the mid-year estimates rolled forward from earlier censuses.

In this volume, the population estimates used for the calculation of fertility rates are the latest consistent estimates available at the time of its production and were published as follows :

Population Estimates by age and sex:
*   2003 estimates; revised 2001 and 2002 estimates - published on 9 September 2004;
*   1992 to 2000 revised estimates - published on 7 October 2004;

Population Estimates by marital status:
*   2003 marital status estimates - published on 4 November 2004;
*   1991 to 2000 marital status estimates; revised 2001 and 2002 marital status estimates - published on 7 October 2004;

These estimates incorporate the results of the local authority population studies, some other corrections and unattributable population change. Further details about the revisions to population estimates can be found at the National Statistics website (www.statistics.gov.uk/popest).

2.2   Occurrences and registrations

Between 1994 and 2000, the cut-off date for inclusion in the annual dataset was births occurring in the reference year registered by 11 February of the following year, this being 42 days after 31 December, the legal time limit for registering a birth. For 2001, the cut-off date was extended to 25 February 2002 to allow increased capture of births registered late. This change means that the annual statistics are prepared on as close to a true occurrences basis as possible, which provides a purer denominator for calculating infant mortality rates.

To avoid artificially inflating the 2001 dataset through the increased capture of late registrations, the start date for the carry over of late registrations from births occurring in 2000 was similarly moved by two weeks.

The total number of births recorded in this 2003 volume includes:

a) births occurring in 2003 which were registered by 25 February 2004, and
b) births occurring in 2002 which were registered between 26 February 2003 and 25 February 2004, that is births in the previous year which had not been tabulated previously.

The number of residual births in (b) was 207 in the 2003 dataset, compared with 519 in the 2000 dataset, reflecting the extended cut-off date.

In the 2000 volume the total number of births included[5]:

a) births which occurred in 2000 registered by 11 February 2001, and
b) births occurring in 1999 which were registered between 12 February 2000 and 11 February 2001, that is births in the previous year which had not been tabulated previously.

Total annual births for 1994 to 1999 were derived in a similar way, except that births for all earlier years were included in the annual totals, not just births in the previous year - see section 1.3 for more details. Up to 1993 the cut-off date was 31 January of the following year, but from 1994 this was then extended to the legal time limit by which a birth should be registered (42 days).

The number of 'residual' births in (b) was about 1,500 to 3,000 per year from 1987 to 1993, but from 1994 it fell to about 500 to 600 annually.

2.3 Areal coverage

The births recorded in this volume are those occurring (and then registered) in England and Wales. No distinction is made between births to civilians and births to non-civilians.

A birth to a mother whose usual residence is outside England and Wales is assigned to the country of residence. These births are included in total figures for England and Wales, but excluded from any sub-division of England and Wales. They are identified as a separate group in **Tables 7.1** to **7.7.**

2.4 Registration of births

Every registrar of births and deaths is required to secure the prompt registration of births occurring within the sub-district covered. Registration of a birth is legally required within 42 days of its occurrence, and the registrar will, if necessary, send a requisition to the person whose duty it is to register the birth.

Under the National Health Service Act 1977, births must also be notified, within 36 hours, to the Director of Public Health in the health authority where the birth occurred. This is carried out by the hospital where the birth took place, or by the midwife or doctor in attendance at the birth. Each month, a list of the births which have occurred in the sub-district is supplied to the registrar, who will then check whether every birth has been registered.

The following people are qualified to give information to the registrar concerning a birth:

a. the mother of the child, and the father if the child was born within marriage;
b. the occupier of the house in which the child was, to the knowledge of that occupier, born;
c. any person present at the birth;
d. any person having charge of the child.

The duty of giving information is placed primarily upon the parents of the child but, in the case of death or inability of the parents, the duty falls on one of the other qualified informants.

The particulars to be registered concerning a birth are prescribed by the Births and Deaths Registration Act 1953, and are covered in section 1.2. Certain other particulars are collected for statistical purposes under the Population Statistics Acts 1938 and 1960, and are not entered in the register. All details are entered on a draft entry *Form 309* (**Annex A**) for a live birth, or *Form 308* (**Annex B**) for a stillbirth. These are checked by the informant before being entered in the register.

The procedures and information required for stillbirths are similar to those for live births. The main difference is the recording of the cause of death of the stillborn child, on evidence given by the doctor or midwife present at the birth, or who examined the body.

Usually, information for the registration of a birth must be given personally by the informant to the registrar for the sub-district in which the birth occurred. However, since April 1997 an informant may supply this information to any registrar by making a declaration of these particulars. The declaration is sent to the registrar of the sub-district where the birth occurred, and that registrar will enter the particulars in the register.

2.5 Visitors and overseas registrations

As noted above, the coverage of this volume is of births occurring, and then registered, in England and Wales. Births to residents of England and Wales which are registered elsewhere are thus excluded, while births registered in England and Wales to mothers whose usual residence is elsewhere, are included. In 2003, there were 218 live births in England and Wales to visitors whose usual residence was elsewhere.

In 2003, 10,440 births occurring outside the United Kingdom to British nationals were voluntarily registered with British Consulates, British High Commissioners, or HM Armed Forces registration centres. Most of these, however, were births to women who had emigrated from the United Kingdom - that is, had lived outside the UK for at least one year - and were thus not residents of England and Wales. Such persons are not included in population estimates for England and Wales.

At any one time some women of childbearing age (defined as age 15-44), usually resident in England and Wales, are temporarily absent overseas. But most of these women are absent for only a short period, and it is unlikely that more than a few hundred per year give birth while overseas. Also, the number of births during 2003 to residents of England and Wales that were registered in Scotland and Northern Ireland were 178 and 71 respectively.

Thus, the number of births to residents of England and Wales occurring outside the country is likely to be of the same order as the number of births occurring in England and Wales to visitors resident elsewhere. The effect on fertility rates of the difference between the definitions used for birth event numerators and population denominators is assumed to be negligible.

2.6 Foundlings

Few, if any, details are known about abandoned children, and they are not included in the statistics given in this volume. However, these infants are included in the Abandoned Children Register maintained at the GRO in Southport. One such entry was made in 2003.

2.7 Country of birth of parents

The country of birth of parents for children born in England and Wales has been recorded at birth registration since April 1969.

Country of birth groupings represent the Commonwealth and European Union as constituted in 2003. The details for country of birth groupings are shown in **Table A**.

Table A   Country groupings for birthplace of parents

United Kingdom	England, Wales, Scotland, Northern Ireland
Elsewhere in United Kingdom	Channel Islands, Isle of Man, UK (part not stated)
Outside United Kingdom	
Irish Republic	Irish Republic, Ireland (part not stated)
Other European Union countries	Austria, Belgium, Denmark, Finland, France, Germany, Greece, Italy, Luxembourg, Netherlands, Portugal, Spain, Sweden
Rest of Europe	All other European countries, including Turkey, Russia and former Soviet republics
Commonwealth	
Australia, Canada and New Zealand	
New Commonwealth	
Asia	Bangladesh, India, Pakistan
East Africa	Kenya, Malawi, Tanzania, Uganda, Zambia
Southern Africa	Botswana, Lesotho, Namibia, South Africa, Swaziland
Rest of Africa	Cameroon, The Gambia, Ghana, Mauritius, Mozambique, Nigeria, Seychelles, Sierra Leone, Zimbabwe
Caribbean	Anguilla, Antigua and Barbuda, Bahamas, Barbados, Belize, Bermuda, British Virgin Islands, Cayman Islands, Dominica, Grenada, Guyana, Jamaica, Montserrat, St Christopher and Nevis, St Lucia, St Vincent, Trinidad and Tobago, Turks and Caicos Islands
Far East	Brunei, Malaysia, Singapore
Mediterranean	Cyprus, Gibraltar, Malta
Rest of New Commonwealth	Cook Islands, Falkland Islands, Fiji, Kiribati, Maldives, Nauru, Papua New Guinea, Pitcairn Islands, St Helena, Solomon Islands, Sri Lanka, Tonga, Tuvalu, Vanuatu, Western Samoa, British Antarctic Territory, British Indian Ocean Territory
Rest of the World	

Major changes from the listing between 1994 and 2003 by year are:

From 1994 - Namibia and South Africa moved from the *Rest of the World* to *New Commonwealth - Southern Africa*.

From 1995 - Cameroon and Mozambique moved from the *Rest of the World* to *New Commonwealth - Rest of Africa*; Austria, Finland and Sweden moved from *Rest of Europe* to *Other European Union countries*.

From 1997 - Hong Kong moved from *New Commonwealth - Far East* to the *Rest of the World*; Fiji moved from the *Rest of the World* to the *Rest of the New Commonwealth*.

From 2003 - Zimbabwe moved from *New Commonwealth - Rest of Africa* to the *Rest of the World*.

2.8 Stillbirths

In Section 41 of the Births and Deaths Registration Act 1953, a stillbirth is defined as 'a child which has issued forth from its mother after the twenty-eighth week of pregnancy and which did not at any time after being completely expelled from its mother breathe or show other signs of life'. This definition was used up to 30 September 1992.

On 1 October 1992 the Stillbirth (Definition) Act 1992 came into force, altering the above definition of a stillbirth to 24 or more weeks completed gestation. Figures for stillbirths from 1993 are thus not fully comparable with those for previous years. The effect of this change on figures for 1992 is analysed in the annual volume for that year[6].

2.9 True birth order

When a birth is within marriage, information is obtained on the number of the mother's previous children, both live births and stillbirths. This allows determination of the *registration birth order* - that is, the number of previous live births plus the birth which has just occurred, counting only those births fathered by a previous or current husband. However, this measure is deficient for fertility statistics in two respects:

a.   at registration, the question on previous live births and stillbirths is not asked where the birth occurred outside marriage, and

b.   at the registration of births and stillbirths occurring within marriage, previous live births occurring outside marriage and where the woman had never been married to the father are not counted. However, because of the ambiguous nature of the question (see **Annex A**) it is possible that births outside marriage where the woman subsequently married the father may not always be included.

The proportion of births occurring outside marriage has risen steadily in recent years. To allow for this, the information collected on birth order has been supplemented to give estimates of overall or *true birth order* - that is, a measure which includes births both within and outside marriage. These estimates are obtained from details provided by the General Household Survey (GHS).

The following example of a hypothetical birth history helps to illustrate the relationship between true birth order, marital birth order and birth order collected at registration:

Birth history	True birth order	Registration birth order	Marital birth order
First birth while cohabiting with man A	1	Not recorded	Not applicable
Second birth while married to man B	2	1	1
Third birth while cohabiting with man C	3	Not recorded	Not applicable
Fourth birth after marriage to man C	4	3	2

Previous volumes of *Birth statistics* used information from the GHS for 1986-1989 to produce the estimates of true birth order shown in **Table 1.7**, **10.3** and **10.5**. From the 2001 volume, GHS information has been used from the surveys for 1986-96, 1998 and 2000 and as a result of this, and an adjustment for the order of multiple births, the figures have changed from those in previous volumes. A description of how the estimates were made is available in a *Population Trends* article[7]. A further revision to the figures resulted from the use of revised mid-year population estimates for 1992-2000, based on results from the 2001 Census, as described in section 2.1.

Table 1.7 shows mean ages for each true birth order calculated in two ways:

  (a) Unstandardised means use only numbers of true order births and single year of age of mother.
  (b) Standardised means use rates per 1,000 female population by true birth order and single year of age of mother. This serves to eliminate the effect of year to year changes in the age-structure of the female population.

2.10 Births within marriage, and sole and joint registration

A birth within marriage is that of a child born to parents who were lawfully married to each other either:

a. at the date of the child's birth, or

b. when the child was conceived, even if they later divorced or the father died before the child's birth.

Only for a birth within marriage will the registrar enter on the draft entry in the register (*Form 309* - **Annex A** or *Form 308* - **Annex B**) confidential particulars relating to the date of the parents' marriage, whether the mother has been married more than once, and the number of the mother's previous live-born and stillborn children - see section 1.2.

Births occurring outside marriage may be registered either jointly or solely. A joint registration records details of both parents, and requires them both to be present. A sole registration records only the mother's details. In a few cases a joint registration is made in the absence of the father if an affiliation order or statutory declaration is provided.

Information from the birth registrations is used to determine whether the mother and father jointly registering a birth outside marriage were usually resident at the same address at the time of registration - see **Table 3.10**. Births with both parents at the same address are identified by a single entry for the informant's usual address, while different addresses are identified by two entries.

2.11 Rates

In this volume, fertility rates have been calculated using the most up-to-date consistent mid-year estimates of the female population, based on the 2001 Census. See section 2.1 and **Appendix Table 1**.

The most commonly used rates are described below[8]:

Crude birth rate

This is the simplest overall measure of fertility in the population, given by the number of live births in a year per 1,000 mid-year population. It is unsophisticated since it takes no account of the composition of the population, in particular the age and sex distribution.

It is given by $(B/P) \times 1,000$
where   B = total live births in the year, and
         P = mid-year population.

In this volume it is used in **Tables 1.1b** and **1.3**.

General fertility rate (GFR)

This is an easily calculated measure of current fertility levels, and denotes the number of live births per 1,000 women aged 15-44. However, it makes no allowance for different sized cohorts of women at childbearing ages.

It is given by (B/P^f_{15-44}) x 1,000

where B = total live births in the year, and

P^f_{15-44}= female population aged 15-44.

In this volume it is used in **Tables 1.1b**, **7.1** and **7.2**.

Age-specific fertility rates (ASFRs)

ASFRs are a measure of fertility specific to the age of the mother, and are useful for comparing the reproductive behaviour of women at different ages. They are calculated by dividing the number of live births to mothers of each age group by the number of females in the population of that age and then expressed per 1,000 women in the age group. They can be calculated for single ages, but are usually calculated for five-year age groups in the reproductive age range, from under 20 up to 45 and over. They provide the basis for a detailed analysis of fertility levels by age of mother when giving birth.

The ASFR based on five-year age groups is given by

$$F_a = (B_a/P^f_a) \times 1,000$$

where F_a = age-specific birth rate for age-group a,

B_a = live births to women in age-group a,

P^f_a = female population in age-group a, and

a = age-group under 20, ... , 45 and over.

For the groups under 20 and 45 and over, the female populations used are women aged 15-19, and women aged 45-49 respectively.

ASFRs are used in **Tables 7.3** and **7.4** of this volume.

Total fertility rate (TFR)

In this volume, for most tables the TFR is derived by summing single-year age-specific fertility rates over all ages within the childbearing lifespan. It is a measure independent of variations in the age distribution of women of childbearing age. It may be interpreted as representing the completed fertility of a synthetic cohort of women - that is, the average number of live children that a woman would bear if the female population experienced the age-specific fertility rates of the calendar year in question throughout their childbearing lifespan.

From the above the TFR = $\sum_{a=\text{under }16}^{a=44\text{ and over}} F_a$

where $F_a = B_a/P^f_a$

and B_a = live births to women in age-group a,

P^f_a = female population in age-group a, and

a = ages under 16, 16, 17, ... , 42, 43, 44 and over.

For the groups under 16 and 44 and over, the female populations used are women aged 15, and women aged 44 respectively.

The TFR is used in **Tables 1.4, 1.9, 2.1, 2.3, 7.1, 7.2** and **9.5.**

Gross reproduction rate (GRR)

The GRR is the sum of age-specific fertility rates for female births only, calculated in the same way as the TFR. It represents the average number of live daughters that a woman would bear in her life if the female population experienced current age-specific fertility rates based on female births throughout their childbearing period.

The GRR is used in **Table 1.4.**

Net reproduction rate (NRR)

The NRR is similar to the GRR, but allows for the effect of mortality to women of childbearing age. Not all women survive to the end of the possible reproductive period. It represents the average number of live daughters that a woman would bear if the female population experienced current age-specific fertility rates and female survival rates throughout their childbearing period.

The NRR is used in **Table 1.4.**

Average family size

Average family size is presented in this volume for women by year of birth and age, in tables on cohort fertility. For each cohort (i.e. women born in a particular year) it represents the number of births per woman, and is shown by birth order for births within marriage. Thus in **Table 10.4** the cohort born in 1951 had, between ages 20 and 24, borne on average 0.65 children per woman. Further, 0.59 were within marriage, 0.31 were first births per woman, 0.21 second births, and so on.

Stillbirth rate

The stillbirth rate is defined as the number of stillbirths per 1,000 live births and stillbirths, and is used in **Tables 1.2, 7.6** and **8.3.**

Sex ratio

Expressed as males per 1,000 females, most often for live births, but also for stillbirths. It is used here in **Tables 1.1b, 1.2** and **4.3.**

Other rates used in this volume include:

Live births within marriage per 1,000 married women, by age - **Tables 1.1b** and **3.1b.**

Live births within marriage per 1,000 married men, by age - **Tables 3.3** and **3.5.**

Live births outside marriage per 1,000 single, widowed and divorced women, by age - **Tables 1.1b** and **3.1b.**

Live births outside marriage per 1,000 live births - **Tables 1.1b, 7.1** and **7.2.**

Paternities within marriage per 1,000 married men, by age - **Table 3.5.**

Stillbirths within marriage per 1,000 married men, by age - **Table 3.5.**

Maternities with multiple births per 1,000 total maternities, by age - **Table 6.1b.**

Maternities within marriage with multiple births, per 1,000 maternities within marriage, by age - **Tables 6.1b** and **6.3.**

Maternities outside marriage with multiple births, per 1,000 maternities outside marriage, by age - **Table 6.1b.**

2.12 Accuracy of information

The accuracy of information contained in the draft birth entry (*Form 309* for a live birth or *Form 308* for a stillbirth - **Annex A** and **B** respectively) is the responsibility of the informant(s) - usually the mother, or the mother and father where the registration is a joint one outside marriage. Wilfully supplying false information may render the informant(s) liable to prosecution for perjury.

It is believed that in general the information supplied by the informant(s) is correct. Computerised internal consistency checks are applied to each record to eliminate, as far as possible, errors made in the supply and recording of information on births.

There are a few, very small, known errors in the database each year which it is not possible to correct. Their effects on the statistics are explained in relevant sections of these notes - for instance, duration of marriage in section 3.4.

2.13 Historical information

The formal registration of live births commenced on 1 July 1837, while stillbirths have been registered only since 1 July 1927. Confidential particulars for statistical purposes have been ascertained since 1 July 1938, under the Population Statistics Act of that year. From the later date, it has also been possible to routinely distinguish multiple births.

The Population Statistics Act 1960, effective from 1 January 1961, added a question on father's date of birth to the confidential particulars requested in the case of births within marriage. This applied also to births outside marriage where the father's name is entered in the register. Questions on father's and mother's place of birth were introduced on 1 April 1969 by the Registration of Births, Deaths and Marriages Regulations 1968.

As noted in section 2.8, the Stillbirth (Definition) Act of 1992 altered the definition of the gestation period for a stillbirth from 28 to 24 completed weeks.

2.14 Legislation

The main statutes concerning birth registration and provision of information on births are given below:

- **Census Act 1920,** which in Section 5 provides for the collection and publication of statistical information on the population by the Registrar General.

- **Population Statistics Act 1938,** which deals with the statistical information collected at registration.

- **Births and Deaths Registration Act 1953,** which covers all aspects of the registration of births and deaths.

- **Registration Service Act 1953,** which in Section 19 requires the Registrar General to provide annual abstracts of live births and stillbirths.

- **Population Statistics Act 1960,** which makes further provision for collecting statistical detail at registration.

- **Abortions Act 1967,** which permits termination of pregnancy by a registered practitioner, subject to certain conditions.

- **Registration of Births, Deaths and Marriages Regulations 1968,** which added questions on father's and mother's place of birth to the details requested at registration.

- **National Health Service Act 1977,** which requires notification of a birth to the health authority where the birth occurred.

- **Stillbirth (Definition) Act 1992,** which altered the definition of a stillbirth to 24 or more weeks completed gestation, instead of the previous definition of 28 or more weeks completed gestation.

- **Health Act 1999,** a section of which includes specific provision for the supply of information on individual births to the National Health Service.

2.15 Symbols and conventions

In this volume:

:	denotes 'not appropriate'
0	denotes less than 0.5
-	denotes nil.

Where data are not yet available, cells in tables are left blank. Rates and percentages calculated from fewer than 20 events (for example **Table 3.5**) are distinguished by italic type as a warning to users that their reliability as a measure may be affected by the small number of events.

Figures in some tables in this publication may not add up due to rounding.

2.16 Further information

Requests for births data, as well as background information on this volume, on unpublished VS tables, and on data quality, should be made to:

Vital Statistics Outputs Branch
Office for National Statistics
Segensworth Road
Titchfield
Fareham
Hants PO15 5RR
Telephone: 01329 813758; Fax: 01329 813548
email: vsob@ons.gov.uk

3 Notes on tables

3.1 Seasonality *(Tables 2.1 to 2.5)*

Seasonally adjusted numbers of live births are obtained using the X-11 ARIMA package developed by Statistics Canada. This software adjusts the monthly totals to eliminate seasonal fluctuations present in the latest 12 years of data, controlling them to the observed annual totals. The annual TFRs shown in **Tables 2.1** and **2.3** have been calculated using monthly and quarterly totals (adjusted for the number of days in those periods). The population denominators are calculated for each month and quarter by apportioning the difference between one mid-year estimate and the next according to the time interval from the first. So the mid-November 2000 estimate is (mid-2001) – (mid-2000) x 4.5/12 + (mid-2000) because mid-November is 4.5 months away from end-June. This method has been used since the 2001 volume.

3.2 Age of parents *(Tables 3.1 to 3.10)*

The mother's or father's date of birth is recorded and translated into the age at the birthday **preceding** the date of the child's birth. This age is often termed age last birthday. Special checks are carried out on those dates of birth which imply that the age of the mother is over 50 years. While most dates of birth are confirmed, these extreme values tend to occur more often among women born outside England and Wales - see **Tables 9.1** to **9.6** - and may result from age misreporting.

If either the mother's date of birth or the father's date of birth (when applicable) is not given, an age is imputed from the last processed record with completely stated and otherwise matching particulars. In 2003, the mother's date of birth was not stated for 0.55 per cent of all live births, and the father's date of birth was not stated for 0.60 per cent of all live births where father's details were present.

3.3 Previous live-born children *(Tables 4.1 to 4.3)*

Information on previous live-born children is available only for women having a birth within marriage. It denotes the number of previous live-born children by the present and any former husband, as stated at registration. This information is also used to determine true birth order - see section 2.9.

If the number of previous live-born children is not given, a value is imputed from the last processed record with completely stated and otherwise matching particulars. In

2003, 0.02 per cent of all live births within marriage had this variable imputed. Of all stillbirths within marriage, 28.05 per cent had this variable imputed.

3.4 Duration of marriage *(Tables 5.1 to 5.3)*

Pre-maritally conceived live births are, by convention, taken to be those where the calculated duration of marriage is up to and including seven completed months. At registration only the month and year of marriage are recorded, so the calculation relates to the interval in completed months between the middle of the month of marriage and the date of the child's birth. Other durations of marriage are calculated in a similar way.

If the date of marriage is not given, a value for the date of marriage is imputed from the last processed record with completely stated and otherwise matching particulars. In 2003, the year of marriage was not stated for 0.74 per cent of all live births within marriage. For women who have been married more than once, duration refers to the length of the current marriage.

In **Table 1.8**, the percentages shown are based on live births where the mother married in the year shown, and the birth was within eight months of marriage, and on first marriages in the same year. Thus the percentages (by age) for year of marriage 2002 would be based on (a) the number of live births in 2002 or 2003 where the mother conceived in 2002 and gave birth within eight months of marriage, and (b) the number of first marriages occurring in 2002.

3.5 Multiple births *(Tables 6.1 to 6.4)*

Multiple births arising from a single pregnancy are counted as one maternity or paternity, although each child born is reckoned separately in analyses of birth statistics. In tables analysing births by the number of previous live-born children, multiple births are counted as if they had occurred separately - in the case of twins for instance, as one first and one second birth.

3.6 Birthweight *(Tables 7.5 and 7.7)*

Birthweight is measured in grams, and is notified to the local health authority by the hospital where the birth took place, or by the midwife or doctor in attendance at the birth. These details are then supplied to the registrar. For stillbirths, details of the weight of the foetus are supplied on a certificate or notification by a doctor or midwife. The certificate or notification is then taken by an informant to the registrar.

In cases where no birthweight is recorded, the birth is included in the total 'all weights' but not distributed amongst the individual categories. (These categories may thus not add to the total in **Tables 7.5** and **7.7.**). In 2003, birthweight was not stated for 0.15 per cent of all live births, and for 2.37 per cent of stillbirths.

3.7 Place of confinement and area of occurrence
(Tables 8.1 to 8.3)

Place of confinement is categorised as follows:

NHS Establishments - generally hospitals, maternity units and maternity wings;

Non-NHS Establishments - including private maternity units, military hospitals, and private hospitals;

At home - denoting the usual place of residence of the mother;

Elsewhere - including all locations not covered above: most of these are at a private residence not that of the mother, or are on the way to a hospital.

A birth is usually assigned to an area according to the usual residence of the mother at the time of birth, as stated at registration. However, a birth may take place in an area other than that of the mother's usual residence. **Table 8.2** shows whether a confinement takes place in the same area as the mother's usual area of residence, or if it occurred in a different area. Births which take place at home or elsewhere are not allocated a health area of occurrence. **Table 8.3** shows the numbers of maternities, live births and stillbirths occurring in NHS and non-NHS establishments, by area of occurrence.

3.8 Country of birth of mother *(Tables 9.1 to 9.6)*

For children born in England and Wales, the country of birth of parents has been recorded at birth registration since April 1969. However, it should be noted that birthplace does not necessarily equate with ethnic group. A fuller discussion of this subject can be found elsewhere[9].

Country of birth groupings represent the Commonwealth and European Union as constituted in 2003 - see section 2.7. Data in **Tables 9.1** and **9.6** have been revised back to 1994.

3.9 Birth cohorts *(Tables 10.1 to 10.5)*

Birth statistics analysed by year of occurrence and by age

of mother have been available since 1938. **Tables 10.1** to **10.5** show these statistics in cohort form - by the year of birth of the mother rather than the year of birth of the child. The years of birth shown are by necessity approximate, since data are available only by calendar year of occurrence and age of mother at childbirth. For instance, women aged 32 giving birth to children in 2003 could have been born in either 1970 or 1971; for convenience, however, such women are here regarded as belonging to the 1971 cohort.

Tables 10.1 to **10.5** all refer to age in completed years. **Table 10.2** gives for a particular cohort (women born in a given year) the average number of live-born children after *n* completed years of age. When data become available for a given cohort from 15 to 45 completed years, then the figure shown after 45 completed years of age measures the average completed family size for women born in that cohort. This is calculated by summing the age-specific birth rates for that cohort shown in **Table 10.1** up to and including age *n*. For example, **Table 10.2** shows that women born in 1970 had given birth to 0.22 children on average after 20 completed years of age. This was calculated by adding the age-specific birth rates for the 1970 cohort in **Table 10.1** up to and including age 20 - that is, $(3+12+28+48+61+72)/1,000 = 0.22$.

3.10 Socio-economic classification as defined by occupation *(Tables 11.1 to 11.5)*

The information on occupation of the father is coded for only a sample of one in ten births. Combining this with the employment status, a code for socio-economic classification (or social class in previous volumes) may be derived. From 1991 to 2000 the occupation of the father was coded using the Standard Occupational Classification SOC90[10], and occupation codes were allocated to the Registrar General's Social Class.

The Standard Occupational Classification is revised every ten years and in 2001 SOC2000[11, 12] replaced SOC90. The coding of employment status also changed in 2001 to be consistent with the 2001 Census and SOC2000. Since 2001, the National Statistics Socio-economic Classification (NS-SEC)[13] has categorised the socio-economic classification of people, and has replaced the Registrar General's Social Class and the Socio-economic Group (SEG). SOC2000 and employment status are used to derive NS-SEC for births. The new classification is based not on skills but on employment conditions, which are now considered to be central to describing the socio-economic structure of modern societies.

NS-SEC has eight analytic classes, the first of which can be subdivided:

1   Higher managerial and professional occupations
        1.1 Large employers and higher managerial occupations
        1.2 Higher professional occupations

2   Lower managerial and professional occupations
3   Intermediate occupations
4   Small employers and own-account workers
5   Lower supervisory and technical occupations
6   Semi-routine occupations
7   Routine occupations
8   Never worked and long-term unemployed

Students, occupations not stated or inadequately described, and occupations not classifiable for other reasons are added as 'Not Classified'.

The sample figures in **Tables 11.1** to **11.5** have been grossed-up to agree with known totals derived from the 100 per cent processing of birth registrations by mother's age and previous live-born children in **Table 4.1a.** This ensures consistency with sub-totals, and improves the quality of sample estimates. **Appendix Tables 3** and **4** show standard errors for selected numbers of births and percentages. Thus, if the estimated grossed-up number in a particular category was 50,000, then from **Appendix Table 3** the standard error of that estimate would be approximately 640. Based on statistical theory, this means that for the type of distribution being considered there is about a 95 per cent chance that the 'true' number in the population lies within two standard errors of the estimates. This true number is that which would have been obtained had all the information been collected, rather than a one in ten sample.

In this example, the 95 per cent confidence interval would be:

50,000 ± 1,300, or 48,700 to 51,300.

In other words, we could say that we are 95 per cent confident that the true value, if we had collected all the information instead of a 10 per cent sample, lies somewhere between 48,700 and 51,300.

3.11 Birth intervals *(Table 11.3)*

Figures in **Table 11.3** showing median birth intervals are produced using two separate sources. The median interval between marriage and first birth is derived from births registration data. It relates to births occurring in England and Wales within marriage only and to the first child that is fathered by the present or any former husband. For remarried women, the interval is measured from the date of the current marriage. Where the first maternity is a multiple birth, the interval is measured from marriage for each resulting child.

The other part of the table shows the median intervals between first, second, third and fourth births. It is derived by the Inland Revenue from a 5 per cent sample of new claims for child benefit from all births occurring in the United Kingdom - whether within or outside marriage. A zero interval is assumed for births resulting from a multiple maternity.

References

1. OPCS (1987). *Birth statistics 1837-1983*, series FM1 no 13.
2. ONS (1999). *Birth statistics* 1998, series FM1 no 27.
3. ONS (2000). *Birth statistics* 1999, series FM1, no 28.
4. ONS (2004). *Key population and vital statistics - local and health authority areas 2002,* series VS no 29, PP1 no 25.
5. ONS (2001) *Birth statistics 2000,* series FM1 no 29.
6. OPCS (1994). *Birth statistics* 1992, series FM1 no 21.
7. Smallwood, S. (2002). 'New estimates of trends in birth order in England and Wales'. *Population Trends* 108, pp 32-48.
8. Shryock, HS. and Siegel, JS. (1973). *The methods and materials of demography,* chapter 16. (US Government Printing Office, Washington DC, 1973).
9. Shaw, C. (1988). 'Components of growth in the ethnic minority population'. *Population Trends* 52, pp 26-30.
10. OPCS (1990). *Standard Occupational Classification: volumes 1-3,* HMSO: London.
11. ONS (2000). *Standard Occupational Classification 2000: Volume 1. Structure and descriptions of unit groups,* TSO: London.
12. ONS (2000). *Standard Occupational Classification 2000: Volume 2. The coding index,* TSO: London.
13. Rose, D and O'Reilly, K (1998). *The ESRC Review of Government Social Classifications,* ESRC & ONS: Swindon.

Glossary

Abortion	The legal termination of a pregnancy under the 1967 Abortion Act.
Age-Specific Fertility Rate (ASFR)	The number of live births to mothers of a particular age per 1,000 women in that age group. Useful for comparing fertility of women at different ages or women with the same ages in different populations.
Annual Reference Volume	ARVs are yearly publications produced by ONS for a variety of topics (for example, FM1 for Birth Statistics).
Cohort	A specific group of people, in this case, those born during a particular year. Analysis using cohorts considers the experience of that group of people over time.
Conception	A pregnancy of a woman which leads either to a maternity or an abortion.
Crude Birth Rate	The number of live births in a year per 1,000 mid-year population.
First Release	Once on annual dataset from a particular source such as birth registration has been quality assured, its first publication is a press, or first, release, which details the main findings.
General Fertility Rate (GFR)	The number of live births per 1,000 women aged 15-44. Measure of current fertility levels.
General Household Survey (GHS)	The GHS is a continuous survey carried out by ONS, collecting information on a range of topics from people living in private households in Great Britain.
GRO	General Register Office – responsible for ensuring the registration of all births, marriages and deaths that have occurred in England and Wales since 1837 and for maintaining a central archive.
Gross Reproduction Rate (GRR)	The sum of age-specific fertility rates for female births only. The average number of live daughters that a woman would bear in her life, if the female population experienced current ASFRs based on female births throughout their childbearing years.
Health Statistics Quarterly	A quarterly ONS publication that covers mortality and health information, including articles and reports on conceptions.
Informant	The person(s), normally one or both parents, who provide the registrar with the information required at the registration of a birth.
Imputation	A method used to add information to an incomplete birth record, using the details from another similar but complete record.
Joint Registration	A birth outside of marriage registered by both the mother and father of the child. Both parents' details are recorded and both must be present.
Live Birth	A baby showing signs of life at birth.

Maternity	A confinement resulting in the birth of one or more live-born or stillborn children. Therefore, the number of maternities (and paternities) is less than the total number of live births and stillbirths.
Median	Statistical term for the value for which half the data are above and half are below. An alternative measure to the mean.
Multiple Births	A single maternity resulting in two or more births.
National Statistics Code of Practice	The principles and protocols followed and upheld by all those involved in producing National Statistics.
Natural Change	The change to the size of the population due to births and deaths (not taking into account the contribution of migration).
Net Reproduction Rate (NRR)	Similar to the GRR, but also takes into account the effect of mortality to women of childbearing age.
NS-SEC	National Statistics Socio-economic Classification categorises the socio-economic classification of people, and has replaced the Registrar General's Social Class and the Socio-economic Group (SEG).
Occurrences	Births which occur in a calendar year.
ONS	Office for National Statistics.
OPCS	Office for Population Censuses and Surveys - joined with Central Statistical Office in 1996 to become ONS.
Parity	The number of live births a woman has had. A woman who has one child has a parity of one. See Registration Birth Order and True Birth Order.
Place of Confinement	The type of building in which a birth occurs.
Population Statistics Act	This Act makes provision for certain information to be collected at the registration of the birth for statistical use. This information is confidential and is not entered on the register.
Population Trends	A quarterly ONS publication that covers population and demographic information, including articles and reports on births.
Ratio	A measure of the relative size of two variables expressed as a proportion.
Registrar	Statutory officer responsible for the registration of births, deaths and marriages.
Registrar General	Statutory appointment with responsibility for the administration of the registration Acts in England and Wales, and other related functions as specified by the relevant legislation.
Registration Birth Order	The number assigned to a birth based on the number of previous live births to that mother, counting only those births fathered by her current or any previous husband(s).
Registration Officer	Generic term for registrar, superintendent registrar and additional registrars.
Registrations	Births that were registered in a particular accounting period, even though some may have occurred in an earlier accounting period.

Reports	Regular articles in *Population Trends* and *Health Statistics Quarterly* on births and conceptions.
Seasonality	The effect of seasonal fluctuations to monthly and quarterly births figures. Monthly totals are adjusted in **Tables 2.1** and **2.3** to eliminate this effect.
Single Men/Women	Persons who have never been married. Legally they are known as bachelors and spinsters.
Singleton	A live birth or stillbirth which is the sole live birth or stillbirth born of a maternity.
SOC2000	Standard Occupational Classification 2000 is the current occupational classification. SOC2000 codes, details of employment status and size of organisation are required for the derivation of NS-SEC. See NS-SEC.
Sole Registration	A birth outside of marriage registered only by the mother. No information on the father is recorded.
Standard Error	A measure of the sampling variation occurring by chance when only part of the total population has been selected, for example, father's occupation.
Standardised Mean Age	The average age (for example, at birth or marriage) of the population in question calculated to take into account the changing distribution of that population by age and other factors over time. This mean should be used when analysing trends.
Stillbirth	A child that has issued forth from its mother after the 24th week of pregnancy, and that did not at any time after being completely expelled from its mother breathe or show any signs of life.
Superintendent Registrar	Statutory officer with responsibilities relating to marriage and other registration functions, as specified in the relevant legislation.
Total Fertility Rate (TFR)	The sum of the age-specific fertility rates. The average number of live children that a woman would bear if the female population experienced the ASFRs of the calendar year in question throughout their childbearing lifespan.
True Birth Order	The number assigned to a birth based on the number of previous live births to that mother, counting all births inside or outside of marriage.
Unstandardised Mean Age	The average age (for example, at birth or marriage) of the population in question, calculated as the actual average for a particular year. It does not take into account the changing distribution of that population over time. This measure should be used when requiring a mean for particular year.
VSOB	Vital Statistics Output Branch (at ONS).

Table 1.1   Live births: occurrence within/outside marriage and sex, 1993-2003   **England and Wales**
a. numbers

Year	All			Within marriage			Outside marriage		
	Total	Males	Females	**Total**	Males	Females	**Total**	Males	Females
1993	**673,467**	345,835	327,632	**456,919**	234,935	221,984	**216,548**	110,900	105,648
1994	**664,726**	341,321	323,405	**449,190**	230,733	218,457	**215,536**	110,588	104,948
1995	**648,138**	332,188	315,950	**428,189**	219,475	208,714	**219,949**	112,713	107,236
1996	**649,485**	333,490	315,995	**416,822**	214,542	202,280	**232,663**	118,948	113,715
1997	**643,095**	329,577	313,518	**404,873**	207,199	197,674	**238,222**	122,378	115,844
1998	**635,901**	325,903	309,998	**395,290**	202,762	192,528	**240,611**	123,141	117,470
1999	**621,872**	319,255	302,617	**379,983**	194,935	185,048	**241,889**	124,320	117,569
2000	**604,441**	309,625	294,816	**365,836**	187,367	178,469	**238,605**	122,258	116,347
2001	**594,634**	304,635	289,999	**356,548**	182,899	173,649	**238,086**	121,736	116,350
2002	**596,122**	306,063	290,059	**354,090**	181,452	172,638	**242,032**	124,611	117,421
2003	**621,469**	318,428	303,041	**364,244**	186,730	177,514	**257,225**	131,698	125,527

Table 1.1   Live births: occurrence within/outside marriage and sex, 1993-2003   **England and Wales**
b. rates and sex ratios

Year	Crude birth rate: all births per 1,000 population of all ages[1]	General fertility rate: all births per 1,000 women aged 15-44[1]	Births within marriage per 1,000 married women aged 15-44[1]	Births outside marriage per 1,000 single, widowed and divorced women aged 15-44[1]	Births outside marriage per 1,000 total births	Sex ratio: male births per 1,000 female births		
						All	Within marriage	Outside marriage
1993	13.2	62.7	84.2	40.7	321.5	1,056	1,058	1,050
1994	13.0	62.0	84.6	39.8	324.2	1,055	1,056	1,054
1995	12.6	60.5	82.7	39.7	339.4	1,051	1,052	1,051
1996	12.6	60.6	82.2	41.2	358.2	1,055	1,061	1,046
1997	12.5	60.0	81.6	41.3	370.4	1,051	1,048	1,056
1998	12.3	59.2	81.3	40.9	378.4	1,051	1,053	1,048
1999	12.0	57.8	79.6	40.4	389.0	1,055	1,053	1,057
2000	11.6	55.9	77.9	39.0	394.8	1,050	1,050	1,051
2001	11.4	54.7	77.3	38.0	400.4	1,050	1,053	1,046
2002	11.3	54.7	78.8	37.7	406.0	1,055	1,051	1,061
2003	11.8	56.8	83.3	39.1	413.9	1,051	1,052	1,049

1 See sections 2.1 and 2.11.

Table 1.2   Stillbirths (numbers, rates and sex ratios):   **England and Wales**
occurrence within/outside marriage and sex, 1993–2003

Year	Numbers									Rates			
	All			Within marriage			Outside marriage			Stillbirths per 1,000 live births and stillbirths	Sex ratio: male births per 1,000 female births		
	Total	Males	Females	**Total**	Males	Females	**Total**	Males	Females		All	Within marriage	Outside marriage
1993	**3,855**	2,076	1,779	**2,431**	1,317	1,114	**1,424**	759	665	5.7	1,167	1,182	1,141
1994	**3,813**	2,034	1,779	**2,361**	1,287	1,074	**1,452**	747	705	5.7	1,143	1,198	1,060
1995	**3,600**	1,912	1,688	**2,224**	1,197	1,027	**1,376**	715	661	5.5	1,133	1,166	1,082
1996	**3,539**	1,808	1,731	**2,114**	1,057	1,057	**1,425**	751	674	5.4	1,044	1,000	1,114
1997	**3,439**	1,801	1,638	**2,010**	1,070	940	**1,429**	731	698	5.3	1,100	1,138	1,047
1998	**3,417**	1,822	1,595	**1,966**	1,054	912	**1,451**	768	683	5.3	1,142	1,156	1,124
1999	**3,305**	1,727	1,578	**1,875**	973	902	**1,430**	754	676	5.3	1,094	1,079	1,115
2000	**3,203**	1,731	1,472	**1,830**	979	851	**1,373**	752	621	5.3	1,176	1,150	1,211
2001	**3,159**	1,725	1,434	**1,771**	936	835	**1,388**	789	599	5.3	1,203	1,121	1,317
2002	**3,372**	1,807	1,565	**1,896**	1,030	866	**1,476**	777	699	5.6	1,155	1,189	1,112
2003	**3,585**	1,874	1,711	**1,961**	1,041	920	**1,624**	833	791	5.7	1,095	1,132	1,053

Table 1.3 Natural change in population (numbers and rates[1]), 1993-2003   **England and Wales**

Year	Numbers			Rates per 1,000 population of all ages		
	Live births	Deaths	Natural change: live births minus deaths	Live births (crude birth rate)	Deaths (crude death rate)	Natural change
1993	673,467	578,799	94,668	13.2	11.4	1.9
1994	664,726	553,194	111,532	13.0	10.8	2.2
1995	648,138	569,683	78,455	12.6	11.1	1.5
1996	649,485	560,135	89,350	12.6	10.9	1.7
1997	643,095	555,281	87,814	12.5	10.8	1.7
1998	635,901	555,015	80,886	12.3	10.7	1.6
1999	621,872	556,118	65,754	12.0	10.7	1.3
2000	604,441	535,664	68,777	11.6	10.3	1.3
2001	594,634	530,373	64,261	11.4	10.1	1.2
2002	596,122	533,527	62,595	11.3	10.1	1.2
2003	621,469	538,254	83,215	11.8	10.2	1.6

1 See sections 2.1 and 2.11.

Table 1.4 Total fertility, gross and net reproduction rates[1], 1993-2003   **England and Wales**

Year	Total fertility rate (TFR)	Gross reproduction rate (GRR)	Net reproduction rate (NRR)
1993	1.76	0.86	0.86
1994	1.75	0.86	0.85
1995	1.72	0.84	0.84
1996	1.74	0.85	0.84
1997	1.73	0.84	0.84
1998	1.72	0.84	0.84
1999	1.70	0.83	0.82
2000	1.65	0.81	0.80
2001	1.63	0.81	0.81
2002	1.65	0.82	0.81
2003	1.73	0.84	0.84

1 See sections 2.1 and 2.11.

Table 1.5 Marriages (numbers and rates[1]): sex, 1992-2002   **England and Wales**

Year	Number of marriages			Marriage rates				
	All	Single men	Single women	Crude marriage rate - all persons marrying per 1,000 population of all ages	Males marrying per 1,000 single, widowed and divorced males aged 16 and over	Females marrying per 1,000 single, widowed and divorced females aged 16 and over	Single men marrying per 1,000 single males aged 16 and over	Single women marrying per 1,000 single females aged 16 and over
1992	311,564	224,152	225,608	12.2	39.6	33.4	37.8	46.3
1993	299,197	213,476	214,987	11.7	37.7	31.8	35.8	43.8
1994	291,069	206,077	206,332	11.4	36.3	30.6	34.3	41.6
1995	283,012	198,208	198,603	11.0	34.7	29.3	32.4	39.3
1996	278,975	193,306	192,707	10.9	33.6	28.5	31.1	37.3
1997	272,536	188,268	188,457	10.6	32.3	27.5	29.7	35.6
1998	267,303	186,329	187,391	10.3	31.1	26.6	28.9	34.7
1999	263,515	184,266	185,328	10.1	30.1	25.8	28.0	33.5
2000	267,961	186,113	187,717	10.3	30.1	25.9	27.7	33.2
2001	249,227	175,721	177,506	9.5	27.4	23.7	25.5	30.6
2002[2]	255,517	179,052	180,605	9.7	27.4	23.9	25.3	30.3

Note: These figures relate only to marriages solemnised in England and Wales.
1 See section 2.1.
2 The figures for 2002 are provisional.

**Table 1.6   Mean age of all women at marriage and   England and Wales
of mothers at live birth, 1993-2003**

Year	Mean age at marriage		Mean age at live birth						
	All brides	Single women	All births	Births outside marriage	Births within marriage				
					All birth orders	First birth	Second birth	Third birth	Fourth birth
1993	29.9	26.2	**28.1**	25.5	29.3	28.0	29.4	30.7	31.8
1994	30.3	26.5	**28.4**	25.8	29.6	28.3	29.7	30.9	32.0
1995	28.1	26.0	**28.5**	26.0	29.8	28.5	30.0	31.1	32.0
1996	31.1	27.2	**28.6**	26.1	30.1	28.8	30.3	31.3	32.2
1997	31.4	27.5	**28.8**	26.2	30.3	29.0	30.5	31.5	32.4
1998	31.6	27.7	**28.9**	26.3	30.5	29.2	30.7	31.8	32.6
1999	31.8	28.0	**29.0**	26.4	30.6	29.3	30.9	32.0	32.7
2000	32.1	28.2	**29.1**	26.5	30.8	29.6	31.1	32.1	32.8
2001	32.2	28.4	**29.2**	26.7	30.9	29.6	31.2	32.2	33.0
2002[1]	32.6	28.7	**29.3**	26.8	31.0	29.7	31.4	32.3	33.1
2003			**29.4**	26.9	31.2	29.9	31.5	32.5	33.1

Note: The mean ages shown in this table are not standardised and therefore take no account of the structure of the population by age, marital status or parity.
1   The mean ages of marriage for 2002 are provisional.

Table 1.7a   Mean age of mother by birth order, 1993-2003   England and Wales

Year	All births	True birth order[1]			
		First	Second	Third	Fourth
1993	**28.1**	26.2	28.5	30.0	31.0
1994	**28.4**	26.5	28.8	30.2	31.2
1995	**28.5**	26.6	29.0	30.3	31.3
1996	**28.6**	26.7	29.2	30.5	31.5
1997	**28.8**	26.8	29.4	30.7	31.7
1998	**28.9**	26.9	29.5	30.8	31.8
1999	**29.0**	26.9	29.7	30.9	31.9
2000	**29.1**	27.1	29.8	31.1	32.0
2001	**29.2**	27.1	29.9	31.1	32.1
2002	**29.3**	27.3	30.0	31.2	32.2
2003	**29.4**	27.4	30.1	31.3	32.2

Note: The mean ages shown in this table are not standardised and therefore take no account of the structure of the population by age, marital status and parity.
1 See section 2.9.

Table 1.7b   Mean age of mother standardised for age[1], 1993-2003   England and Wales

Year	All births	True birth order[2]			
		First	Second	Third	Fourth
1993	**27.9**	25.8	28.5	30.1	31.3
1994	**28.1**	26.0	28.7	30.2	31.4
1995	**28.2**	26.1	28.8	30.3	31.4
1996	**28.2**	26.0	28.8	30.4	31.5
1997	**28.3**	26.1	29.0	30.4	31.6
1998	**28.3**	26.2	29.1	30.5	31.6
1999	**28.4**	26.3	29.1	30.5	31.6
2000	**28.5**	26.4	29.2	30.6	31.7
2001	**28.6**	26.5	29.3	30.7	31.7
2002	**28.7**	26.7	29.4	30.7	31.8
2003	**28.8**	26.9	29.5	30.8	31.8

1 Standardised to take account of the age structure of the population. This measure is more appropriate for use when analysing trends or making comparisons
   between different geographies.
2 See section 2.9.

Table 1.8 Percentage of first marriages with a birth within 8 months of marriage: age of mother at marriage, 1992-2002[1]   **England and Wales**

Year of marriage	Woman's age at marriage				
	Under 20	20-24	25-29	30-44	Under 45
1992	20.1	9.3	8.8	11.3	10.1
1993	22.8	9.9	8.9	11.2	10.6
1994	22.5	9.9	8.3	11.1	10.2
1995	20.6	10.3	8.3	10.8	10.1
1996	23.1	11.0	8.5	10.5	10.5
1997	21.9	11.1	8.6	10.8	10.5
1998	22.7	11.4	8.3	10.4	10.4
1999	22.7	11.1	8.0	10.0	10.0
2000	19.9	10.9	7.5	8.8	9.3
2001	22.0	10.9	7.2	8.7	9.2
2002	22.3	11.0	7.6	8.9	9.4

1 See section 3.4.

Table 1.9 Total fertility rates[1]: occurrence within/outside marriage, birth order[2] and age of mother, 1993-2003   **England and Wales**

Year	All live births	Live births outside marriage	Live births within marriage						Year	All live births	Live births outside marriage	Live births within marriage					
			All	Birth order								All	Birth order				
				First	Second	Third	Fourth	Fifth and later					First	Second	Third	Fourth	Fifth and later
	All ages of mother at birth									**25-29**							
1993	**1.76**	0.59	1.17	0.46	0.43	0.19	0.06	0.04	1993	**0.57**	0.14	0.43	0.19	0.16	0.06	0.02	0.01
1994	**1.75**	0.60	1.15	0.45	0.42	0.18	0.06	0.04	1994	**0.56**	0.14	0.42	0.19	0.15	0.06	0.02	0.01
1995	**1.72**	0.62	1.10	0.44	0.40	0.17	0.06	0.03	1995	**0.54**	0.15	0.39	0.18	0.14	0.05	0.02	0.01
1996	**1.74**	0.66	1.08	0.43	0.39	0.17	0.06	0.03	1996	**0.53**	0.16	0.38	0.17	0.14	0.05	0.02	0.01
1997	**1.73**	0.68	1.05	0.41	0.39	0.16	0.06	0.03	1997	**0.52**	0.16	0.36	0.16	0.13	0.05	0.02	0.01
1998	**1.72**	0.69	1.03	0.41	0.38	0.16	0.05	0.03	1998	**0.51**	0.17	0.34	0.16	0.12	0.04	0.01	0.01
1999	**1.70**	0.70	1.00	0.41	0.36	0.15	0.05	0.03	1999	**0.49**	0.17	0.32	0.15	0.11	0.04	0.01	0.01
2000	**1.65**	0.69	0.97	0.40	0.35	0.14	0.05	0.03	2000	**0.47**	0.17	0.31	0.14	0.11	0.04	0.01	0.01
2001	**1.63**	0.69	0.95	0.39	0.35	0.14	0.05	0.03	2001	**0.46**	0.16	0.29	0.14	0.10	0.04	0.01	0.01
2002	**1.65**	0.70	0.95	0.40	0.35	0.13	0.05	0.03	2002	**0.46**	0.17	0.29	0.14	0.10	0.04	0.01	0.01
2003	**1.73**	0.74	0.99	0.42	0.36	0.14	0.05	0.03	2003	**0.48**	0.18	0.30	0.15	0.10	0.04	0.01	0.01
	Under 20									**30-34**							
1993	**0.15**	0.13	0.02	0.02	0.01	0.00	0.00	0.00	1993	**0.43**	0.08	0.35	0.11	0.14	0.07	0.02	0.01
1994	**0.14**	0.12	0.02	0.02	0.00	0.00	0.00	0.00	1994	**0.44**	0.08	0.36	0.11	0.15	0.07	0.02	0.01
1995	**0.15**	0.13	0.02	0.02	0.00	0.00	0.00	0.00	1995	**0.44**	0.09	0.35	0.11	0.14	0.06	0.02	0.01
1996	**0.15**	0.13	0.02	0.02	0.00	0.00	0.00	0.00	1996	**0.45**	0.10	0.35	0.11	0.14	0.06	0.02	0.01
1997	**0.15**	0.14	0.02	0.01	0.00	0.00	0.00	0.00	1997	**0.45**	0.10	0.35	0.12	0.14	0.06	0.02	0.01
1998	**0.16**	0.14	0.02	0.01	0.00	0.00	0.00	0.00	1998	**0.45**	0.11	0.35	0.12	0.14	0.06	0.02	0.01
1999	**0.15**	0.14	0.02	0.01	0.00	0.00	0.00	0.00	1999	**0.45**	0.11	0.34	0.12	0.14	0.05	0.02	0.01
2000	**0.15**	0.13	0.02	0.01	0.00	0.00	0.00	0.00	2000	**0.44**	0.11	0.33	0.12	0.13	0.05	0.02	0.01
2001	**0.14**	0.13	0.02	0.01	0.00	0.00	0.00	0.00	2001	**0.44**	0.11	0.33	0.12	0.13	0.05	0.02	0.01
2002	**0.14**	0.12	0.02	0.01	0.00	0.00	0.00	0.00	2002	**0.45**	0.12	0.34	0.13	0.13	0.05	0.02	0.01
2003	**0.14**	0.12	0.01	0.01	0.00	0.00	0.00	0.00	2003	**0.48**	0.13	0.35	0.14	0.14	0.05	0.02	0.01
	20-24									**35 and over**							
1993	**0.41**	0.20	0.21	0.11	0.07	0.02	0.00	0.00	1993	**0.20**	0.04	0.16	0.04	0.05	0.04	0.02	0.02
1994	**0.39**	0.20	0.19	0.10	0.07	0.02	0.00	0.00	1994	**0.21**	0.05	0.16	0.04	0.06	0.04	0.02	0.02
1995	**0.38**	0.21	0.17	0.09	0.06	0.02	0.00	0.00	1995	**0.21**	0.05	0.17	0.04	0.06	0.04	0.02	0.02
1996	**0.38**	0.22	0.16	0.09	0.06	0.02	0.00	0.00	1996	**0.22**	0.05	0.17	0.04	0.06	0.04	0.02	0.02
1997	**0.38**	0.22	0.15	0.08	0.05	0.02	0.00	0.00	1997	**0.23**	0.06	0.17	0.04	0.06	0.04	0.02	0.02
1998	**0.37**	0.22	0.15	0.08	0.05	0.02	0.00	0.00	1998	**0.24**	0.06	0.18	0.04	0.06	0.04	0.02	0.01
1999	**0.37**	0.22	0.14	0.08	0.05	0.01	0.00	0.00	1999	**0.24**	0.06	0.18	0.05	0.06	0.04	0.02	0.01
2000	**0.35**	0.22	0.13	0.07	0.05	0.01	0.00	0.00	2000	**0.25**	0.07	0.18	0.05	0.07	0.04	0.02	0.01
2001	**0.35**	0.22	0.13	0.07	0.04	0.01	0.00	0.00	2001	**0.25**	0.07	0.18	0.05	0.07	0.04	0.02	0.01
2002	**0.35**	0.22	0.13	0.07	0.04	0.01	0.00	0.00	2002	**0.26**	0.07	0.19	0.05	0.07	0.04	0.02	0.01
2003	**0.36**	0.23	0.13	0.07	0.04	0.01	0.00	0.00	2003	**0.28**	0.08	0.20	0.06	0.08	0.04	0.02	0.01

1 See sections 2.1 and 2.11.
2 See section 2.9.

Table 2.1 Live births (actual and seasonally adjusted numbers and rates[1]): quarter of occurrence, 1993-2003 — **England and Wales**

Year	Total	Quarter ending				Total	Quarter ending			
		31 March	30 June	30 September	31 December		31 March	30 June	30 September	31 December
	Number (thousands)					Total fertility rate (TFR)				
	Live births					Actual TFR				
1993	**673.5**	162.3	170.1	175.9	165.2	**1.76**	1.72	1.78	1.83	1.72
1994	**664.7**	164.3	170.7	168.9	160.9	**1.75**	1.75	1.80	1.77	1.69
1995	**648.1**	158.5	164.7	167.4	157.5	**1.72**	1.70	1.75	1.76	1.66
1996	**649.5**	157.3	158.1	169.9	164.2	**1.74**	1.69	1.69	1.81	1.76
1997	**643.1**	158.1	163.3	164.9	156.8	**1.73**	1.72	1.76	1.76	1.68
1998	**635.9**	155.8	158.6	166.1	155.4	**1.72**	1.71	1.72	1.79	1.68
1999	**621.9**	152.1	157.3	160.1	152.4	**1.70**	1.68	1.71	1.73	1.65
2000	**604.4**	148.7	150.7	155.0	150.1	**1.65**	1.63	1.65	1.69	1.64
2001	**594.6**	145.5	148.8	153.0	147.4	**1.63**	1.62	1.64	1.67	1.61
2002	**596.1**	143.3	147.2	155.0	150.6	**1.65**	1.60	1.63	1.70	1.66
2003	**621.5**	147.4	155.2	162.9	156.0	**1.73**	1.66	1.73	1.80	1.72
	Seasonally adjusted live births[2]					Seasonally adjusted TFR[2]				
1993		167.3	167.9	168.9	168.5		1.75	1.76	1.78	1.77
1994		168.9	168.8	164.5	163.6		1.77	1.77	1.72	1.73
1995		162.7	163.2	162.8	160.2		1.72	1.73	1.72	1.71
1996		160.1	158.0	162.9	166.7		1.71	1.68	1.76	1.79
1997		162.6	163.1	160.6	158.3		1.75	1.75	1.72	1.71
1998		160.4	158.6	159.2	156.9		1.74	1.71	1.74	1.71
1999		156.3	157.2	156.3	153.7		1.70	1.71	1.69	1.68
2000		151.6	151.3	151.1	151.4		1.66	1.65	1.64	1.66
2001		150.0	149.1	148.3	147.5		1.65	1.63	1.63	1.63
2002		148.3	147.6	148.2	150.8		1.64	1.63	1.66	1.67
2003		152.8	155.7	156.8	155.9		1.70	1.73	1.75	1.74

Note: Figures may not add exactly due to rounding - see section 2.15.
1 See sections 2.1 and 2.11.
2 See section 3.1.

Table 2.2 Stillbirths: quarter of occurrence, 1993-2003 — **England and Wales**

Year	Total	Quarter ending			
		31 March	30 June	30 September	31 December
1993	**3,855**	941	944	981	989
1994	**3,813**	986	926	952	949
1995	**3,600**	958	865	904	873
1996	**3,539**	875	886	917	861
1997	**3,439**	867	870	839	863
1998	**3,417**	865	849	824	879
1999	**3,305**	867	834	818	786
2000	**3,203**	808	776	797	822
2001	**3,159**	785	799	745	830
2002	**3,372**	827	869	849	827
2003	**3,585**	867	911	933	874

Table 2.3 Live births (actual and seasonally adjusted numbers and rates[1]): month of occurrence, 1993-2003   **England and Wales**

Year	Total	January	February	March	April	May	June	July	August	Sept-ember	Oct-ober	Nov-ember	Dec-ember
Numbers (thousands)													
Live births													
1993	**673.5**	55.4	50.7	56.3	54.8	57.6	57.7	59.0	57.8	59.0	56.8	52.6	55.8
1994	**664.7**	55.4	51.0	57.9	55.4	58.1	57.1	57.3	55.5	56.1	55.5	52.2	53.1
1995	**648.1**	53.5	49.7	55.4	52.2	56.7	55.8	56.5	55.6	55.3	54.9	51.4	51.2
1996	**649.5**	53.5	50.4	53.4	50.5	53.8	53.7	57.6	56.0	56.4	56.2	53.8	54.2
1997	**643.1**	54.5	49.5	54.1	54.0	55.4	53.9	56.4	54.8	53.7	53.0	50.7	53.1
1998	**635.9**	53.4	48.8	53.6	52.4	53.0	53.1	56.4	54.4	55.3	53.6	50.1	51.6
1999	**621.9**	51.2	47.6	53.3	50.8	53.5	53.0	54.5	52.9	52.7	51.0	49.9	51.6
2000	**604.4**	50.5	47.0	51.1	49.0	51.8	49.9	52.6	51.9	50.4	50.9	49.8	49.5
2001	**594.6**	50.6	45.2	49.8	47.8	51.5	49.5	51.3	51.0	50.7	51.4	48.4	47.6
2002	**596.1**	49.2	45.2	49.0	47.9	50.7	48.5	52.0	51.1	51.9	52.2	48.6	49.7
2003	**621.5**	50.6	46.0	50.8	50.8	52.9	51.5	55.4	53.3	54.1	54.2	50.6	51.3
Seasonally adjusted live births[2]													
1993		56.5	55.5	55.3	55.8	55.9	56.2	56.2	56.5	57.0	56.6	55.2	56.7
1994		56.2	55.9	56.8	56.7	56.5	55.6	55.0	53.9	54.5	55.0	54.4	54.2
1995		54.2	54.4	54.2	53.9	54.7	54.7	54.1	54.1	53.8	54.5	53.2	52.5
1996		53.6	53.4	53.1	52.0	52.5	53.5	54.5	54.9	55.3	55.5	55.9	55.4
1997		54.5	54.3	53.7	55.1	54.4	53.6	53.2	53.8	52.2	52.1	52.8	53.3
1998		53.6	53.5	53.3	53.3	52.4	52.8	53.2	53.3	53.6	53.0	51.9	51.9
1999		51.7	52.2	52.4	52.0	52.6	52.6	51.9	51.8	51.0	50.6	51.2	51.9
2000		51.0	50.1	50.6	50.8	50.4	50.1	50.6	50.4	49.2	50.2	50.7	50.5
2001		50.5	49.9	49.7	49.4	50.0	49.7	49.3	49.3	49.5	49.6	49.3	48.5
2002		49.1	49.9	49.3	49.2	49.2	49.2	49.3	49.7	50.5	50.4	49.6	50.8
2003		50.6	51.0	51.2	51.9	51.8	52.1	52.5	52.2	52.3	52.1	52.0	51.8
Total fertility rates (TFR)													
Actual TFR													
1993	**1.76**	1.71	1.73	1.73	1.74	1.77	1.84	1.82	1.79	1.89	1.76	1.68	1.73
1994	**1.75**	1.71	1.75	1.79	1.77	1.80	1.83	1.78	1.72	1.80	1.72	1.68	1.66
1995	**1.72**	1.67	1.71	1.72	1.68	1.76	1.80	1.76	1.74	1.79	1.72	1.67	1.61
1996	**1.74**	1.69	1.70	1.68	1.64	1.69	1.75	1.82	1.77	1.84	1.78	1.76	1.72
1997	**1.73**	1.73	1.74	1.71	1.76	1.75	1.76	1.79	1.74	1.76	1.68	1.67	1.69
1998	**1.72**	1.70	1.72	1.71	1.72	1.69	1.75	1.80	1.74	1.83	1.72	1.66	1.66
1999	**1.70**	1.64	1.69	1.71	1.68	1.71	1.75	1.75	1.70	1.75	1.64	1.66	1.66
2000	**1.65**	1.63	1.62	1.65	1.63	1.67	1.66	1.70	1.68	1.69	1.65	1.67	1.61
2001	**1.63**	1.64	1.62	1.61	1.59	1.66	1.65	1.66	1.65	1.70	1.67	1.63	1.55
2002	**1.65**	1.60	1.62	1.59	1.61	1.65	1.63	1.69	1.67	1.75	1.70	1.64	1.63
2003	**1.73**	1.65	1.50	1.66	1.66	1.73	1.68	1.81	1.75	1.77	1.77	1.66	1.68
Seasonally adjusted TFR[2]													
1993		1.77	1.74	1.73	1.75	1.75	1.76	1.77	1.78	1.80	1.78	1.74	1.79
1994		1.77	1.76	1.79	1.79	1.78	1.75	1.74	1.70	1.73	1.74	1.72	1.72
1995		1.72	1.73	1.72	1.71	1.73	1.74	1.72	1.72	1.72	1.74	1.70	1.68
1996		1.72	1.71	1.70	1.66	1.68	1.71	1.75	1.76	1.78	1.79	1.80	1.79
1997		1.76	1.75	1.73	1.78	1.75	1.73	1.72	1.74	1.69	1.69	1.71	1.73
1998		1.74	1.74	1.73	1.73	1.70	1.71	1.73	1.74	1.75	1.73	1.70	1.70
1999		1.69	1.71	1.71	1.70	1.71	1.72	1.70	1.70	1.67	1.66	1.68	1.71
2000		1.67	1.64	1.65	1.66	1.65	1.64	1.66	1.66	1.62	1.65	1.67	1.67
2001		1.66	1.64	1.63	1.63	1.64	1.64	1.62	1.63	1.64	1.64	1.63	1.61
2002		1.63	1.65	1.63	1.63	1.63	1.63	1.63	1.65	1.68	1.67	1.65	1.69
2003		1.69	1.70	1.70	1.73	1.72	1.73	1.75	1.74	1.74	1.74	1.74	1.73

Note: Figures may not add exactly due to rounding - see section 2.15.
1 See sections 2.1 and 2.11.
2 See section 3.1.

Table 2.4 Maternities, live births and stillbirths: quarter and month of occurrence[1], within/outside marriage and sex, 2003   **England and Wales**

Quarter/month of occurrence	Maternities	Live births Total	Within marriage	Outside marriage	Stillbirths Total	Within marriage	Outside marriage	Live births Male	Female	Stillbirths Male	Female
Annual Total	**615,787**	**621,469**	**364,244**	**257,225**	**3,585**	**1,961**	**1,624**	**318,428**	**303,041**	**1,874**	**1,711**
March quarter	146,092	**147,441**	86,427	61,014	**867**	481	386	75,647	71,794	437	430
June quarter	153,755	**155,152**	92,371	62,781	**911**	482	429	79,678	75,474	485	426
September quarter	161,430	**162,860**	95,243	67,617	**933**	490	443	83,480	79,380	494	439
December quarter	154,510	**156,016**	90,203	65,813	**874**	508	366	79,623	76,393	458	416
January	50,150	**50,579**	29,415	21,164	**319**	178	141	25,839	24,740	162	157
February	45,574	**46,017**	26,890	19,127	**252**	144	108	23,583	22,434	134	118
March	50,368	**50,845**	30,122	20,723	**296**	159	137	26,225	24,620	141	155
April	50,308	**50,761**	30,284	20,477	**289**	161	128	25,916	24,845	143	146
May	52,425	**52,910**	31,516	21,394	**315**	148	167	27,237	25,673	180	135
June	51,022	**51,481**	30,571	20,910	**307**	173	134	26,525	24,956	162	145
July	54,881	**55,425**	32,678	22,747	**313**	166	147	28,548	26,877	160	153
August	52,875	**53,311**	31,008	22,303	**319**	166	153	27,135	26,176	176	143
September	53,674	**54,124**	31,557	22,567	**301**	158	143	27,797	26,327	158	143
October	53,696	**54,159**	31,659	22,500	**317**	190	127	27,643	26,516	163	154
November	50,099	**50,599**	29,358	21,241	**277**	157	120	25,803	24,796	146	131
December	50,715	**51,258**	29,186	22,072	**280**	161	119	26,177	25,081	149	131

1 Including a small number of births which occurred in 2002 and were registered after the 'cut-off' date in 2003 - see section 2.2.

Table 2.5 Live birth occurrences[1] in 2003: quarter and month of occurrence and month of registration   **England and Wales**

Quarter/month of occurrence	January	February	March	April	May	June	July	August	September	October	November	December	Registrations from 1.1.04 to 25.2.04	Total
Annual Total	**25,502**	**44,485**	**49,229**	**48,777**	**51,942**	**52,745**	**56,360**	**49,213**	**55,194**	**56,471**	**48,020**	**46,475**	**37,056**	**621,469**
March	25,502	44,439	49,138	25,802	2,411	97	25	10	5	1	3	4	4	**147,441**
June	-	-	4	22,947	49,522	52,644	28,199	1,703	95	22	7	3	6	**155,152**
September	-	-	8	5	-	1	28,130	47,500	55,091	29,696	2,272	125	32	**162,860**
December	-	46	79	23	9	3	6	-	3	26,752	45,738	46,343	37,014	**156,016**
January	25,502	22,674	2,299	79	7	2	6	3	1	-	2	1	3	**50,579**
February	-	21,765	22,491	1,679	60	15	5	-	1	-	-	1	-	**46,017**
March	-	-	24,348	24,044	2,344	80	14	7	3	1	1	2	1	**50,845**
April	-	-	-	22,943	25,531	2,193	70	12	3	4	2	1	2	**50,761**
May	-	-	-	1	23,991	26,537	2,302	57	13	6	1	-	2	**52,910**
June	-	-	4	3	-	23,914	25,827	1,634	79	12	4	2	2	**51,481**
July	-	-	1	2	-	-	28,129	24,540	2,648	87	11	4	3	**55,425**
August	-	-	2	2	-	1	-	22,960	27,813	2,428	81	18	6	**53,311**
September	-	-	5	1	-	-	1	-	24,630	27,181	2,180	103	23	**54,124**
October	-	-	4	1	1	-	1	-	-	26,749	24,209	3,012	182	**54,159**
November	-	4	9	7	2	3	-	-	2	-	21,529	25,533	3,510	**50,599**
December	-	42	66	15	6	-	5	-	1	3	-	17,798	33,322	**51,258**

1 Including a small number of births which occurred in 2002 and were registered after the 'cut-off' date in 2003 - see section 2.2.

Table 3.1 Live births: age of mother and occurrence within/outside marriage, 1993-2003   **England and Wales**
a. numbers

Year	Age of mother at birth							
	All ages	Under 20	20-24	25-29	30-34	35-39	40-44	45 and over
	All live births							
1993	**673,467**	45,121	151,975	235,961	171,061	58,824	9,986	539
1994	**664,726**	42,026	140,240	229,102	179,568	63,061	10,241	488
1995	**648,138**	41,938	130,744	217,418	181,202	65,517	10,779	540
1996	**649,485**	44,667	125,732	211,103	186,377	69,503	11,516	587
1997	**643,095**	46,372	118,589	202,792	187,528	74,900	12,332	582
1998	**635,901**	48,285	113,537	193,144	188,499	78,881	12,980	575
1999	**621,872**	48,375	110,722	181,931	185,311	81,281	13,617	635
2000	**604,441**	45,846	107,741	170,701	180,113	84,974	14,403	663
2001	**594,634**	44,189	108,844	159,926	178,920	86,495	15,499	761
2002	**596,122**	43,467	110,959	153,379	180,532	90,449	16,441	895
2003	**621,469**	44,236	116,622	156,931	187,214	97,386	18,205	875
	Live births within marriage							
1993	**456,919**	6,875	76,950	178,456	139,671	46,919	7,621	427
1994	**449,190**	6,099	69,227	170,605	145,563	49,668	7,662	366
1995	**428,189**	5,623	61,029	157,855	144,200	51,129	7,944	409
1996	**416,822**	5,365	54,651	148,770	145,898	53,265	8,421	452
1997	**404,873**	5,233	49,068	139,383	145,293	56,671	8,797	428
1998	**395,290**	5,278	45,724	130,747	144,599	59,320	9,189	433
1999	**379,983**	5,333	43,190	120,716	140,330	60,470	9,466	478
2000	**365,836**	4,742	40,262	111,606	136,165	62,671	9,910	480
2001	**356,548**	4,640	40,736	103,131	133,710	63,202	10,582	547
2002	**354,090**	4,582	40,712	97,583	134,093	65,369	11,110	641
2003	**364,244**	4,338	40,887	98,694	138,002	69,595	12,113	615
	Live births outside marriage							
1993	**216,548**	38,246	75,025	57,505	31,390	11,905	2,365	112
1994	**215,536**	35,927	71,013	58,497	34,005	13,393	2,579	122
1995	**219,949**	36,315	69,715	59,563	37,002	14,388	2,835	131
1996	**232,663**	39,302	71,081	62,333	40,479	16,238	3,095	135
1997	**238,222**	41,139	69,521	63,409	42,235	18,229	3,535	154
1998	**240,611**	43,007	67,813	62,397	43,900	19,561	3,791	142
1999	**241,889**	43,042	67,532	61,215	44,981	20,811	4,151	157
2000	**238,605**	41,104	67,479	59,095	43,948	22,303	4,493	183
2001	**238,086**	39,549	68,108	56,795	45,210	23,293	4,917	214
2002	**242,032**	38,885	70,247	55,796	46,439	25,080	5,331	254
2003	**257,225**	39,898	75,735	58,237	49,212	27,791	6,092	260

Table 3.1 Live births: age of mother and occurrence within/outside marriage, 1993-2003 **England and Wales**
  b. rates[1]

Year	Age of mother at birth							
	All ages	Under 20	20-24	25-29	30-34	35-39	40-44	45 and over
	All live births per 1,000 women[2]							
1993	**62.7**	30.9	82.5	114.4	87.4	34.1	5.9	0.3
1994	**62.0**	28.9	79.0	112.2	89.4	35.8	6.1	0.3
1995	**60.5**	28.5	76.4	108.4	88.3	36.3	6.5	0.3
1996	**60.6**	29.7	77.0	106.6	89.8	37.5	6.9	0.3
1997	**60.0**	30.2	76.0	104.3	89.8	39.4	7.3	0.3
1998	**59.2**	30.9	74.9	101.5	90.6	40.4	7.5	0.3
1999	**57.8**	30.9	73.0	98.3	89.6	40.6	7.7	0.4
2000	**55.9**	29.3	70.0	94.3	87.9	41.4	8.0	0.4
2001	**54.7**	28.0	69.0	91.7	88.0	41.5	8.4	0.5
2002	**54.7**	27.0	69.2	91.6	89.8	43.0	8.6	0.5
2003	**56.8**	26.8	71.2	96.4	94.8	46.4	9.3	0.5
	Live births within marriage per 1,000 married women[2]							
1993	**84.2**	290.2	196.3	167.1	104.7	36.3	5.8	0.3
1994	**84.6**	283.1	200.7	166.9	108.9	38.3	6.0	0.3
1995	**82.7**	270.7	203.4	164.0	108.4	39.2	6.3	0.3
1996	**82.2**	258.7	210.5	164.2	110.9	40.4	6.7	0.3
1997	**81.6**	257.7	218.2	165.1	112.4	42.6	7.0	0.3
1998	**81.3**	260.4	227.4	167.0	114.8	44.2	7.3	0.3
1999	**79.6**	269.9	230.0	166.5	114.8	44.7	7.5	0.4
2000	**77.9**	268.8	224.2	164.8	115.2	46.0	7.8	0.4
2001	**77.3**	287.1	228.9	164.9	117.0	46.6	8.2	0.5
2002	**78.8**	352.4	245.1	172.0	122.5	48.7	8.5	0.5
2003	**83.3**	370.2	254.3	188.5	132.4	52.9	9.2	0.5
	Live births outside marriage per 1,000 single, widowed and divorced women[2]							
1993	**40.7**	26.6	51.7	57.8	50.5	27.5	6.2	0.3
1994	**39.8**	25.1	49.7	57.4	50.6	28.9	6.5	0.3
1995	**39.7**	25.0	49.4	57.2	51.3	28.9	6.9	0.3
1996	**41.2**	26.5	51.8	58.0	53.3	30.4	7.3	0.3
1997	**41.3**	27.2	52.0	57.6	53.1	31.9	7.9	0.4
1998	**40.9**	27.9	51.5	55.7	53.5	32.1	8.0	0.3
1999	**40.4**	27.8	50.8	54.4	53.1	32.0	8.4	0.4
2000	**39.0**	26.6	49.6	52.2	50.7	32.3	8.5	0.4
2001	**38.0**	25.3	48.7	50.7	50.7	32.1	8.8	0.5
2002	**37.7**	24.4	48.8	50.4	50.7	33.0	8.9	0.5
2003	**39.1**	24.4	51.3	52.8	52.8	35.3	9.6	0.5

1 The rates for women of all ages, under 20 and 45 and over are based upon the population of women aged 15-44, 15-19 and 45-49 respectively.
2 See sections 2.1 and 2.11.

Table 3.2 Maternities (total), live births and stillbirths (total and female): age of mother and occurrence within/outside marriage, 2003

Age of mother at birth	Maternities			Births			
	Total	Within marriage	Outside marriage	Live		Still	
				Total	Female	Total	Female
All ages	615,787	360,114	255,673	621,469	303,041	3,585	1,711
11	2	-	2	2	-	-	-
12	9	-	9	7	2	2	2
13	26	-	26	26	12	-	-
14	187	3	184	186	91	2	1
15	1,034	10	1,024	1,025	500	12	5
16	3,635	98	3,537	3,621	1,746	31	12
17	8,706	416	8,290	8,679	4,209	65	27
18	13,416	1,148	12,268	13,441	6,521	87	47
19	17,230	2,671	14,559	17,249	8,408	109	54
Under 20	44,245	4,346	39,899	44,236	21,489	308	148
20	19,878	4,303	15,575	19,889	9,580	124	67
21	21,670	6,200	15,470	21,748	10,538	109	44
22	23,519	7,971	15,548	23,609	11,591	125	51
23	25,398	10,374	15,024	25,507	12,428	139	75
24	25,682	11,836	13,846	25,869	12,643	112	46
20-24	116,147	40,684	75,463	116,622	56,780	609	283
25	26,117	13,913	12,204	26,230	12,829	176	84
26	28,104	16,427	11,677	28,290	14,041	133	69
27	31,107	19,293	11,814	31,329	15,157	195	89
28	33,889	22,552	11,337	34,187	16,681	169	84
29	36,585	25,746	10,839	36,895	18,012	231	111
25-29	155,802	97,931	57,871	156,931	76,720	904	437
30	38,562	27,761	10,801	38,957	18,900	225	109
31	39,829	29,247	10,582	40,275	19,531	219	104
32	39,273	29,143	10,130	39,760	19,397	245	111
33	35,211	26,342	8,869	35,669	17,434	197	82
34	32,060	23,688	8,372	32,553	15,942	162	83
30-34	184,935	136,181	48,754	187,214	91,204	1,048	489
35	27,832	20,322	7,510	28,300	13,838	148	75
36	23,198	16,773	6,425	23,557	11,379	114	56
37	18,943	13,414	5,529	19,271	9,421	123	54
38	14,796	10,422	4,374	15,012	7,271	100	51
39	11,072	7,522	3,550	11,246	5,532	76	46
35-39	95,841	68,453	27,388	97,386	47,441	561	282
40	7,591	5,107	2,484	7,690	3,770	68	32
41	4,938	3,260	1,678	4,984	2,515	40	16
42	2,950	1,934	1,016	2,977	1,435	18	11
43	1,694	1,091	603	1,713	843	18	9
44	825	554	271	841	407	4	2
40-44	17,998	11,946	6,052	18,205	8,970	148	70
45	389	262	127	401	199	4	2
46	192	127	65	204	110	2	-
47	84	63	21	95	37	-	-
48	54	39	15	58	31	-	-
49	34	28	6	39	15	1	-
45-49	753	519	234	797	392	7	2
50 and over	66	54	12	78	45	-	-

England and Wales

| Births within marriage | | | | Births outside marriage | | | | Age of mother at birth |
| Live | | Still | | Live | | Still | | |
Total	Female	Total	Female	Total	Female	Total	Female	
364,244	**177,514**	**1,961**	**920**	**257,225**	**125,527**	**1,624**	**791**	**All ages**
-	-	-	-	2	-	-	-	11
-	-	-	-	7	2	2	2	12
-	-	-	-	26	12	-	-	13
3	2	-	-	183	89	2	1	14
10	4	-	-	1,015	496	12	5	15
97	41	1	1	3,524	1,705	30	11	16
409	193	7	4	8,270	4,016	58	23	17
1,149	572	8	5	12,292	5,949	79	42	18
2,670	1,360	20	11	14,579	7,048	89	43	19
4,338	**2,172**	**36**	**21**	**39,898**	**19,317**	**272**	**127**	**Under 20**
4,304	2,091	28	15	15,585	7,489	96	52	20
6,227	3,059	27	12	15,521	7,479	82	32	21
8,014	3,901	41	18	15,595	7,690	84	33	22
10,423	5,070	45	19	15,084	7,358	94	56	23
11,919	5,810	53	18	13,950	6,833	59	28	24
40,887	**19,931**	**194**	**82**	**75,735**	**36,849**	**415**	**201**	**20-24**
13,972	6,847	91	43	12,258	5,982	85	41	25
16,536	8,237	68	36	11,754	5,804	65	33	26
19,443	9,383	115	51	11,886	5,774	80	38	27
22,766	11,099	103	53	11,421	5,582	66	31	28
25,977	12,667	158	65	10,918	5,345	73	46	29
98,694	**48,233**	**535**	**248**	**58,237**	**28,487**	**369**	**189**	**25-29**
28,088	13,608	144	71	10,869	5,292	81	38	30
29,603	14,392	141	66	10,672	5,139	78	38	31
29,526	14,316	176	76	10,234	5,081	69	35	32
26,713	13,013	142	59	8,956	4,421	55	23	33
24,072	11,779	112	57	8,481	4,163	50	26	34
138,002	**67,108**	**715**	**329**	**49,212**	**24,096**	**333**	**160**	**30-34**
20,670	10,095	103	53	7,630	3,743	45	22	35
17,057	8,194	83	41	6,500	3,185	31	15	36
13,653	6,698	82	37	5,618	2,723	41	17	37
10,567	5,102	69	37	4,445	2,169	31	14	38
7,648	3,710	42	26	3,598	1,822	34	20	39
69,595	**33,799**	**379**	**194**	**27,791**	**13,642**	**182**	**88**	**35-39**
5,192	2,555	44	21	2,498	1,215	24	11	40
3,289	1,646	29	11	1,695	869	11	5	41
1,959	965	12	7	1,018	470	6	4	42
1,107	541	11	5	606	302	7	4	43
566	269	2	1	275	138	2	1	44
12,113	**5,976**	**98**	**45**	**6,092**	**2,994**	**50**	**25**	**40-44**
272	134	3	1	129	65	1	1	45
136	67	1	-	68	43	1	-	46
71	26	-	-	24	11	-	-	47
42	22	-	-	16	9	-	-	48
32	14	-	-	7	1	1	-	49
553	**263**	**4**	**1**	**244**	**129**	**3**	**1**	**45-49**
62	32	-	-	16	13	-	-	50 and over

Table 3.3 Live births within marriage (numbers and rates[1]): age of father, 1993-2003 **England and Wales**

Year	Age of father at birth											
	All ages	Under 20	20-24	25-29	30-34	35-39	40-44	45-49	50-54	55-59	60-64	65 and over
	Numbers											
1993	**456,919**	1,363	37,948	140,674	159,845	77,470	26,673	8,584	2,645	1,071	420	226
1994	**449,190**	1,156	33,372	132,213	161,509	80,723	27,362	8,721	2,540	1,047	376	171
1995	**428,189**	1,015	28,873	120,070	157,518	80,946	26,934	8,826	2,475	1,008	364	160
1996	**416,822**	867	25,013	112,012	155,022	83,279	27,819	8,775	2,584	933	347	171
1997	**404,873**	940	21,895	103,320	151,162	86,097	28,579	8,684	2,794	894	350	158
1998	**395,290**	964	19,755	96,415	148,267	87,684	29,345	8,701	2,807	820	368	164
1999	**379,983**	1,028	18,703	87,896	141,629	88,329	29,986	8,328	2,756	878	304	146
2000	**365,836**	907	16,762	79,351	135,810	89,607	30,782	8,446	2,861	843	330	137
2001	**356,548**	907	16,872	72,668	132,570	88,657	31,981	8,756	2,775	863	344	155
2002	**354,090**	907	16,827	68,390	130,835	90,871	33,337	8,858	2,676	920	320	149
2003	**364,244**	835	16,753	66,854	133,951	95,381	36,257	9,858	2,839	1,002	313	201
	Rates: live births within marriage per 1,000 married men by age											
1993	**39.7**	207.8	202.8	177.8	137.5	64.2	21.0	6.2	2.4	1.0	0.4	0.1
1994	**39.3**	191.4	208.2	176.8	139.5	67.0	22.2	6.3	2.2	1.0	0.4	0.1
1995	**37.7**	169.1	210.8	171.7	136.9	66.6	22.3	6.4	2.1	0.9	0.4	0.1
1996	**36.9**	145.2	214.6	172.3	136.6	67.8	23.2	6.4	2.2	0.9	0.4	0.1
1997	**36.0**	155.1	222.1	173.6	136.1	69.7	23.9	6.7	2.2	0.9	0.4	0.1
1998	**35.3**	155.8	229.7	177.2	137.5	70.7	24.4	7.0	2.1	0.8	0.4	0.1
1999	**34.1**	164.4	238.3	176.8	135.7	70.7	24.8	6.9	2.0	0.8	0.3	0.1
2000	**32.9**	153.3	225.2	173.0	134.9	71.4	25.1	7.1	2.1	0.8	0.3	0.1
2001	**32.2**	165.1	227.0	173.1	136.5	70.9	25.7	7.4	2.1	0.7	0.4	0.1
2002	**32.1**	206.6	242.3	184.1	142.0	73.7	26.6	7.5	2.1	0.7	0.3	0.1
2003	**33.3**	221.0	244.4	198.1	154.3	79.0	28.7	8.4	2.4	0.8	0.3	0.1

1 See sections 2.1 and 2.11.

Table 3.4 Paternities (total), live births and stillbirths (total and male): **England and Wales**
age of father and occurrence within/outside marriage, 2003

Age of father at birth	Paternities within marriage	Births within marriage				Births outside marriage registered[1] on joint information of parents			
		Live		Still		Live		Still	
		Total	Male	Total	Male	Total	Male	Total	Male
All ages	360,114	364,244	186,730	1,961	1,041	212,350	108,957	1,259	648
13	-	-	-	-	-	3	2	-	-
14	-	-	-	-	-	16	10	1	-
15	2	2	-	-	-	141	78	-	-
16	7	7	4	-	-	591	285	5	1
17	39	39	25	-	-	1,636	812	9	6
18	218	215	117	4	1	3,327	1,776	26	15
19	573	572	286	4	2	5,483	2,865	32	14
Under 20	**839**	**835**	**432**	**8**	**3**	**11,197**	**5,828**	**73**	**36**
20	1,144	1,143	609	10	6	7,177	3,706	47	20
21	2,064	2,073	1,096	17	7	8,604	4,476	48	30
22	3,120	3,129	1,564	20	14	9,651	4,916	59	27
23	4,416	4,428	2,243	27	19	10,364	5,480	71	41
24	5,962	5,980	3,086	42	20	10,019	5,144	62	24
20-24	**16,706**	**16,753**	**8,598**	**116**	**66**	**45,815**	**23,722**	**287**	**142**
25	7,783	7,794	3,995	47	16	9,759	5,029	33	21
26	9,851	9,900	5,011	63	32	9,816	5,048	53	29
27	12,759	12,877	6,504	63	35	10,190	5,261	59	23
28	16,155	16,281	8,367	86	53	10,643	5,474	76	37
29	19,847	20,002	10,301	94	62	10,654	5,509	62	33
25-29	**66,395**	**66,854**	**34,178**	**353**	**198**	**51,062**	**26,321**	**283**	**143**
30	23,641	23,878	12,230	103	50	10,779	5,466	50	25
31	26,750	27,022	13,815	135	72	10,894	5,656	57	33
32	28,281	28,585	14,740	149	85	10,476	5,318	64	28
33	27,415	27,771	14,288	142	79	9,416	4,806	56	30
34	26,354	26,695	13,811	135	80	8,978	4,625	57	35
30-34	**132,441**	**133,951**	**68,884**	**664**	**366**	**50,543**	**25,871**	**284**	**151**
35	24,134	24,443	12,534	130	70	8,203	4,289	50	22
36	21,552	21,879	11,259	118	48	7,484	3,772	36	17
37	18,753	18,994	9,758	103	48	6,749	3,410	36	21
38	16,125	16,371	8,276	81	42	5,755	2,915	31	17
39	13,489	13,694	6,962	76	37	4,847	2,447	29	14
35-39	**94,053**	**95,381**	**48,789**	**508**	**245**	**33,038**	**16,833**	**182**	**91**
40	10,904	11,059	5,675	69	34	4,160	2,022	30	16
41	8,553	8,668	4,442	55	28	3,392	1,722	25	13
42	6,867	6,994	3,607	31	15	2,739	1,395	18	13
43	5,325	5,394	2,776	36	14	2,243	1,122	13	8
44	4,070	4,142	2,076	27	13	1,642	837	8	6
40-44	**35,719**	**36,257**	**18,576**	**218**	**104**	**14,176**	**7,098**	**94**	**56**
45	3,150	3,204	1,652	23	15	1,325	685	11	6
46	2,372	2,410	1,216	13	10	1,032	517	7	3
47	1,783	1,804	938	14	9	833	413	4	2
48	1,369	1,389	722	6	2	646	342	3	2
49	1,029	1,051	560	6	3	549	286	5	2
45-49	**9,703**	**9,858**	**5,088**	**62**	**39**	**4,385**	**2,243**	**30**	**15**
50-54	2,778	2,839	1,431	20	13	1,438	707	14	9
55-59	973	1,002	484	8	6	497	242	10	3
60-64	310	313	158	2	1	131	58	-	-
65-69	129	131	79	2	-	50	25	2	2
70-74	40	42	17	-	-	13	6	-	-
75 and over	28	28	16	-	-	5	3	-	-

1 44,875 live births and 365 stillbirths occurred outside marriage and were registered by the mother alone. For these sole registrations only the mother's details are recorded.

Table 3.5   Rates[1] of paternities within marriage by age of father, and rates[1] of   **England and Wales**
live births and stillbirths within marriage by age of father, 2003

| Age of father at birth | Paternities within marriage per 1,000 married men | Births within marriage per 1,000 married men | | | |
| | | Live | | Still | |
		Total	Male	Total	Male
All ages	**32.9**	**33.3**	**17.07**	**0.18**	**0.10**
under 20	222.0	221.0	114.32	*2.12*	*0.79*
20-24	243.7	244.4	125.44	1.69	0.96
25-29	196.8	198.1	101.29	1.05	0.59
30-34	152.5	154.3	79.33	0.76	0.42
35-39	77.9	79.0	40.40	0.42	0.20
40-44	28.3	28.7	14.71	0.17	0.08
45-49	8.2	8.4	4.32	0.05	0.03
50-54	2.3	2.4	1.19	0.02	*0.01*
55-59	0.8	0.8	0.38	*0.01*	0.00
60-64	0.3	0.3	0.16	*0.00*	0.00
65 and over	0.1	0.1	0.04	*0.00*	-

1   See sections 2.1 and 2.11.

Table 3.6   Live births within marriage: **England and Wales**
age of mother and of father, 2003

| Age of father at birth | Age of mother at birth | | | | | | | | |
	All ages	Under 20	20-24	25-29	30-34	35-39	40-44	45-49	50 and over
All ages	**364,244**	**4,338**	**40,887**	**98,694**	**138,002**	**69,595**	**12,113**	**553**	**62**
Under 20	**835**	457	292	53	25	6	1	-	1
20-24	**16,753**	1,898	10,621	3,186	796	210	42	-	-
25-29	**66,854**	1,293	17,322	34,780	11,300	1,921	235	2	1
30-34	**133,951**	480	8,893	42,130	68,238	13,074	1,102	33	1
35-39	**95,381**	133	2,593	13,572	42,736	33,212	3,064	60	11
40-44	**36,257**	47	781	3,511	11,246	15,709	4,815	140	8
45-49	**9,858**	13	251	950	2,577	3,903	1,967	189	8
50-54	**2,839**	5	89	312	718	1,023	588	86	18
55-59	**1,002**	7	29	126	235	349	222	27	7
60-64	**313**	3	6	42	80	123	47	7	5
65 and over	**201**	2	10	32	51	65	30	9	2

Table 3.7   Stillbirths within marriage: **England and Wales**
age of mother and of father, 2003

| Age of father at birth | Age of mother at birth | | | | | | | | |
	All ages	Under 20	20-24	25-29	30-34	35-39	40-44	45-49	50 and over
All ages	**1,961**	**36**	**194**	**535**	**715**	**379**	**98**	**4**	**-**
Under 20	**8**	5	2	1	-	-	-	-	-
20-24	**116**	17	57	33	5	4	-	-	-
25-29	**353**	8	84	191	52	14	4	-	-
30-34	**664**	4	36	203	344	67	10	-	-
35-39	**508**	1	8	79	241	158	21	-	-
40-44	**218**	1	5	21	54	95	41	1	-
45-49	**62**	-	1	5	8	31	15	2	-
50-54	**20**	-	-	1	8	6	4	1	-
55-59	**8**	-	1	1	1	3	2	-	-
60-64	**2**	-	-	-	-	1	1	-	-
65 and over	**2**	-	-	-	2	-	-	-	-

Table 3.8 Live births outside marriage: age of mother and of father and whether sole or joint registration, 2003

England and Wales

Age of father at birth	Age of mother at birth							
	All ages	Under 20	20-24	25-29	30-34	35-39	40-44	45 and over
Total live births outside marriage								
	257,225	39,898	75,735	58,237	49,212	27,791	6,092	260
Solely registered								
	44,875	11,192	14,434	8,555	6,094	3,625	927	48
Jointly registered								
All ages	212,350	28,706	61,301	49,682	43,118	24,166	5,165	212
Under 20	11,197	8,468	2,229	348	112	37	3	-
20-24	45,815	13,848	24,522	5,220	1,699	459	66	1
25-29	51,062	4,007	20,229	18,041	6,616	1,883	279	7
30-34	50,543	1,522	9,253	16,142	17,139	5,647	825	15
35-39	33,038	573	3,438	6,770	11,723	9,041	1,442	51
40-44	14,176	207	1,141	2,279	4,191	4,796	1,505	57
45-49	4,385	53	329	621	1,104	1,552	677	49
50-54	1,438	17	97	174	367	519	245	19
55-59	497	3	38	59	112	178	96	11
60 and over	199	8	25	28	55	54	27	2

Note: Figures for jointly registered live births outside marriage include a small number of cases registered by the mother alone, for which the father's name was included in the birth register.

Table 3.9   Live births outside marriage (numbers and percentages):
age of mother and whether sole or joint registration, 1993-2003

England and Wales

Year		Age of mother at birth											
		All ages	Under 20	20-24	25-29	30-34	35 and over	All ages	Under 20	20-24	25-29	30-34	35 and over
		Number						Percentage by age					
1993	Total	216,548	38,246	75,025	57,505	31,390	14,382						
	Sole	50,242	12,535	17,476	11,418	5,967	2,846	23.2	32.8	23.3	19.9	19.0	19.8
	Joint	166,306	25,711	57,549	46,087	25,423	11,536	76.8	67.2	76.7	80.1	81.0	80.2
1994	Total	215,536	35,927	71,013	58,497	34,005	16,094						
	Sole	49,030	12,007	16,408	11,390	6,204	3,021	22.7	33.4	23.1	19.5	18.2	18.8
	Joint	166,506	23,920	54,605	47,107	27,801	13,073	77.3	66.6	76.9	80.5	81.8	81.2
1995	Total	219,949	36,315	69,715	59,563	37,002	17,354						
	Sole	47,916	11,880	15,626	10,864	6,485	3,061	21.8	32.7	22.4	18.2	17.5	17.6
	Joint	172,033	24,435	54,089	48,699	30,517	14,293	78.2	67.3	77.6	81.8	82.5	82.4
1996	Total	232,663	39,302	71,081	62,333	40,479	19,468						
	Sole	51,016	12,946	16,245	11,350	6,978	3,497	21.9	32.9	22.9	18.2	17.2	18.0
	Joint	181,647	26,356	54,836	50,983	33,501	15,971	78.1	67.1	77.1	81.8	82.8	82.0
1997	Total	238,222	41,139	69,521	63,409	42,235	21,918						
	Sole	50,582	15,590	11,015	6,997	3,757	21.2	32.1	22.4	17.4	16.6	17.1
	Joint	187,640	27,916	53,931	52,394	35,238	18,161	78.8	67.9	77.6	82.6	83.4	82.9
1998	Total	240,611	43,007	67,813	62,397	43,900	23,494						
	Sole	49,960	13,830	14,838	10,672	6,768	3,852	20.8	32.2	21.9	17.1	15.4	16.4
	Joint	190,651	29,177	52,975	51,725	37,132	19,642	79.2	67.8	78.1	82.9	84.6	83.6
1999	Total	241,889	43,042	67,532	61,215	44,981	25,119						
	Sole	48,203	13,194	14,477	9,936	6,615	3,981	19.9	30.7	21.4	16.2	14.7	15.8
	Joint	193,686	29,848	53,055	51,279	38,366	21,138	80.1	69.3	78.6	83.8	85.3	84.2
2000	Total	238,605	41,104	67,479	59,095	43,948	26,979						
	Sole	45,773	12,459	13,919	9,213	6,116	4,066	19.2	30.3	20.6	15.6	13.9	15.1
	Joint	192,832	28,645	53,560	49,882	37,832	22,913	80.8	69.7	79.4	84.4	86.1	84.9
2001	Total	238,086	39,549	68,108	56,795	45,210	28,424						
	Sole	43,744	11,669	13,583	8,582	5,875	4,035	18.4	29.5	19.9	15.1	13.0	14.2
	Joint	194,342	27,880	54,525	48,213	39,335	24,389	81.6	70.5	80.1	84.9	87.0	85.8
2002	Total	242,032	38,885	70,247	55,796	46,439	30,665						
	Sole	43,129	11,229	13,599	8,133	5,949	4,219	17.8	28.9	19.4	14.6	12.8	13.8
	Joint	198,903	27,656	56,648	47,663	40,490	26,446	82.2	71.1	80.6	85.4	87.2	86.2
2003	Total	257,225	39,898	75,735	58,237	49,212	34,143						
	Sole	44,875	11,192	14,434	8,555	6,094	4,600	17.4	28.1	19.1	14.7	12.4	13.5
	Joint	212,350	28,706	61,301	49,682	43,118	29,543	82.6	71.9	80.9	85.3	87.6	86.5

Note: Figures for jointly registered live births outside marriage include a small number of cases registered by the mother alone, for which the father's name was included in the birth register.

Table 3.10 Jointly registered live births outside marriage (numbers and percentages): age of mother and whether parents were usually resident at the same or different addresses, 1993-2003

England and Wales

Year	Addresses of mother and father	Age of mother at birth											
		All ages	Under 20	20-24	25-29	30-34	35 and over	All ages	Under 20	20-24	25-29	30-34	35 and over
		Numbers of jointly registered births outside marriage						Percentage by age					
1993	Total	166,306	25,711	57,549	46,087	25,423	11,536						
	Same	118,758	14,134	40,168	35,307	20,084	9,065	71.4	55.0	69.8	76.6	79.0	78.6
	Different	47,548	11,577	17,381	10,780	5,339	2,471	28.6	45.0	30.2	23.4	21.0	21.4
1994	Total	166,506	23,920	54,605	47,107	27,801	13,073						
	Same	123,874	14,168	39,827	37,168	22,288	10,423	74.4	59.2	72.9	78.9	80.2	79.7
	Different	42,632	9,752	14,778	9,939	5,513	2,650	25.6	40.8	27.1	21.1	19.8	20.3
1995	Total	172,033	24,435	54,089	48,699	30,517	14,293						
	Same	127,789	14,424	39,274	38,376	24,376	11,339	74.3	59.0	72.6	78.8	79.9	79.3
	Different	44,244	10,011	14,815	10,323	6,141	2,954	25.7	41.0	27.4	21.2	20.1	20.7
1996	Total	181,647	26,356	54,836	50,983	33,501	15,971						
	Same	135,282	15,410	39,978	40,384	26,875	12,635	74.5	58.5	72.9	79.2	80.2	79.1
	Different	46,365	10,946	14,858	10,599	6,626	3,336	25.5	41.5	27.1	20.8	19.8	20.9
1997	Total	187,640	27,916	53,931	52,394	35,238	18,161						
	Same	141,740	16,551	39,784	42,020	28,715	14,670	75.5	59.3	73.8	80.2	81.5	80.8
	Different	45,900	11,365	14,147	10,374	6,523	3,491	24.5	40.7	26.2	19.8	18.5	19.2
1998	Total	190,651	29,177	52,975	51,725	37,132	19,642						
	Same	146,521	17,589	39,818	42,212	30,938	15,964	76.9	60.3	75.2	81.6	83.3	81.3
	Different	44,130	11,588	13,157	9,513	6,194	3,678	23.1	39.7	24.8	18.4	16.7	18.7
1999	Total	193,686	29,848	53,055	51,279	38,366	21,138						
	Same	149,584	17,833	40,190	42,126	32,152	17,283	77.2	59.7	75.8	82.2	83.8	81.8
	Different	44,102	12,015	12,865	9,153	6,214	3,855	22.8	40.3	24.2	17.8	16.2	18.2
2000	Total	192,832	28,645	53,560	49,882	37,832	22,913						
	Same	149,510	17,011	40,450	41,236	31,795	19,018	77.5	59.4	75.5	82.7	84.0	83.0
	Different	43,322	11,634	13,110	8,646	6,037	3,895	22.5	40.6	24.5	17.3	16.0	17.0
2001	Total	194,342	27,880	54,525	48,213	39,335	24,389						
	Same	150,421	16,215	40,843	39,797	33,306	20,260	77.4	58.2	74.9	82.5	84.7	83.1
	Different	43,921	11,665	13,682	8,416	6,029	4,129	22.6	41.8	25.1	17.5	15.3	16.9
2002	Total	198,903	27,656	56,648	47,663	40,490	26,446						
	Same	154,086	15,824	42,315	39,554	34,326	22,067	77.5	57.2	74.7	83.0	84.8	83.4
	Different	44,817	11,832	14,333	8,109	6,164	4,379	22.5	42.8	25.3	17.0	15.2	16.6
2003	Total	212,350	28,706	61,301	49,682	43,118	29,543						
	Same	163,374	16,000	45,225	40,867	36,630	24,652	76.9	55.7	73.8	82.3	85.0	83.4
	Different	48,976	12,706	16,076	8,815	6,488	4,891	23.1	44.3	26.2	17.7	15.0	16.6

Note: Figures for jointly registered live births outside marriage include a small number of cases registered by the mother alone, for which the father's name was included in the birth register.

Table 4.1   Live births within marriage: number of previous live-born children and age of mother (five-year age groups), 1993-2003   **England and Wales**
a.   all married women

Year	Number of previous live-born children[1]						Year	Number of previous live-born children[1]					
	Total	0	1	2	3	4 or more		Total	0	1	2	3	4 or more
	All ages of mother at birth							**25-29**					
1993	**456,919**	178,141	169,445	71,789	23,781	13,763	1993	**178,456**	77,612	66,682	24,023	7,356	2,783
1994	**449,190**	176,046	166,343	69,735	23,582	13,484	1994	**170,605**	75,660	62,721	22,550	6,854	2,820
1995	**428,189**	168,118	158,102	66,692	22,320	12,957	1995	**157,855**	71,036	57,282	20,513	6,472	2,552
1996	**416,822**	163,020	153,780	65,291	22,038	12,693	1996	**148,770**	67,178	53,445	19,576	6,153	2,418
1997	**404,873**	157,049	150,404	63,220	21,515	12,685	1997	**139,383**	63,139	50,024	18,108	5,793	2,319
1998	**395,290**	155,708	146,850	60,391	20,327	12,014	1998	**130,747**	60,602	46,404	16,375	5,198	2,168
1999	**379,983**	153,423	139,490	56,375	19,517	11,178	1999	**120,716**	57,385	41,763	14,735	4,907	1,926
2000	**365,836**	146,509	134,691	54,920	18,590	11,126	2000	**111,606**	52,673	38,376	14,129	4,594	1,834
2001	**356,548**	143,908	132,228	52,111	17,553	10,748	2001	**103,131**	48,849	35,650	12,750	4,179	1,703
2002	**354,090**	145,241	130,346	50,307	17,560	10,636	2002	**97,583**	47,111	32,985	11,848	4,028	1,611
2003	**364,244**	150,982	132,873	51,981	17,641	10,767	2003	**98,694**	48,439	32,480	12,108	4,072	1,595
	Under 20							**30-34**					
1993	**6,875**	5,196	1,523	142	12	2	1993	**139,671**	42,665	55,939	26,948	9,002	5,117
1994	**6,099**	4,687	1,268	132	9	3	1994	**145,563**	46,110	58,615	26,779	8,989	5,070
1995	**5,623**	4,308	1,187	116	10	2	1995	**144,200**	46,573	58,458	26,098	8,335	4,736
1996	**5,365**	4,223	1,034	102	5	1	1996	**145,898**	47,742	59,084	25,968	8,426	4,678
1997	**5,233**	4,121	1,018	84	9	1	1997	**145,293**	48,090	59,414	25,112	8,116	4,561
1998	**5,278**	4,188	1,001	85	4	-	1998	**144,599**	49,529	58,944	23,998	7,740	4,388
1999	**5,333**	4,290	930	99	13	1	1999	**140,330**	50,031	56,622	22,314	7,293	4,070
2000	**4,742**	3,810	842	81	7	2	2000	**136,165**	49,400	54,760	21,120	6,832	4,053
2001	**4,640**	3,781	771	82	5	1	2001	**133,710**	49,695	53,780	19,828	6,456	3,951
2002	**4,582**	3,817	687	71	5	2	2002	**134,093**	51,026	53,730	18,999	6,538	3,800
2003	**4,338**	3,471	791	67	8	1	2003	**138,002**	54,170	54,343	19,248	6,413	3,828
	20-24							**35 and over**					
1993	**76,950**	40,399	26,827	7,915	1,516	293	1993	**54,967**	12,269	18,474	12,761	5,895	5,568
1994	**69,227**	36,421	23,926	7,115	1,484	281	1994	**57,696**	13,168	19,813	13,159	6,246	5,310
1995	**61,029**	32,340	20,634	6,453	1,367	235	1995	**59,482**	13,861	20,541	13,512	6,136	5,432
1996	**54,651**	28,925	18,481	5,789	1,231	225	1996	**62,138**	14,952	21,736	13,856	6,223	5,371
1997	**49,068**	25,891	16,558	5,257	1,153	209	1997	**65,896**	15,808	23,390	14,659	6,444	5,595
1998	**45,724**	24,303	15,537	4,679	993	212	1998	**68,942**	17,086	24,964	15,254	6,392	5,246
1999	**43,190**	23,474	14,443	4,175	910	188	1999	**70,414**	18,243	25,732	15,052	6,394	4,993
2000	**40,262**	21,571	13,744	3,950	827	170	2000	**73,061**	19,055	26,969	15,640	6,330	5,067
2001	**40,736**	22,188	13,723	3,897	751	177	2001	**74,331**	19,395	28,304	15,554	6,162	4,916
2002	**40,712**	22,432	13,506	3,866	739	169	2002	**77,120**	20,855	29,438	15,523	6,250	5,054
2003	**40,887**	22,187	13,915	3,834	777	174	2003	**82,323**	22,715	31,344	16,724	6,371	5,169

1 See section 2.9.

**Table 4.1 Live births within marriage: number of previous live-born children
and age of mother (five-year age groups), 1993-2003
b. women married once only**

England and Wales

Year	Number of previous live-born children[1]						Year	Number of previous live-born children[1]					
	Total	0	1	2	3	4 or more		Total	0	1	2	3	4 or more
	All ages of mother at birth							**25-29**					
1993	421,059	168,624	157,843	63,587	19,759	11,246	1993	169,746	74,896	63,733	22,159	6,528	2,430
1994	413,994	166,466	154,855	61,954	19,696	11,023	1994	162,526	73,011	60,001	20,925	6,137	2,452
1995	394,904	158,963	147,266	59,295	18,705	10,675	1995	150,644	68,634	54,897	19,020	5,814	2,279
1996	384,240	153,994	143,223	58,094	18,546	10,383	1996	142,375	65,075	51,296	18,258	5,592	2,154
1997	373,428	148,419	140,155	56,425	18,023	10,406	1997	133,555	61,201	48,104	16,953	5,223	2,074
1998	365,105	147,207	136,867	53,797	17,292	9,942	1998	125,642	58,883	44,644	15,357	4,790	1,968
1999	352,494	145,500	130,434	50,644	16,645	9,271	1999	116,462	55,900	40,375	13,890	4,525	1,772
2000	340,054	139,210	126,117	49,451	15,999	9,277	2000	107,887	51,426	37,148	13,379	4,274	1,660
2001	332,607	136,975	124,162	47,145	15,248	9,077	2001	99,996	47,766	34,622	12,146	3,911	1,551
2002	331,265	138,490	122,629	45,760	15,323	9,063	2002	94,901	46,213	32,077	11,340	3,786	1,485
2003	341,645	144,274	125,276	47,325	15,478	9,292	2003	96,292	47,569	31,678	11,656	3,886	1,503
	Under 20							**30-34**					
1993	6,867	5,194	1,521	141	10	1	1993	124,850	38,626	51,012	23,624	7,398	4,190
1994	6,092	4,682	1,266	132	9	3	1994	130,643	41,841	53,612	23,590	7,460	4,140
1995	5,609	4,300	1,183	114	10	2	1995	130,170	42,557	53,710	23,095	6,923	3,885
1996	5,355	4,218	1,031	100	5	1	1996	131,973	43,639	54,427	23,009	7,039	3,859
1997	5,226	4,116	1,016	84	9	1	1997	132,225	44,292	54,982	22,343	6,803	3,805
1998	5,273	4,186	999	85	3	-	1998	132,210	45,773	54,755	21,354	6,634	3,694
1999	5,329	4,287	929	99	13	1	1999	129,006	46,439	52,704	20,112	6,286	3,465
2000	4,737	3,806	841	81	7	2	2000	125,811	46,204	51,177	18,990	5,960	3,480
2001	4,638	3,780	770	82	5	1	2001	124,250	46,644	50,523	17,963	5,669	3,451
2002	4,574	3,813	685	69	5	2	2002	125,244	48,079	50,631	17,381	5,796	3,357
2003	4,331	3,468	788	66	8	1	2003	129,567	51,362	51,430	17,638	5,737	3,400
	20-24							**35 and over**					
1993	75,750	39,994	26,409	7,644	1,442	261	1993	43,846	9,914	15,168	10,019	4,381	4,364
1994	68,262	36,117	23,556	6,897	1,427	265	1994	46,471	10,815	16,420	10,410	4,663	4,163
1995	60,190	32,056	20,331	6,278	1,309	216	1995	48,291	11,416	17,145	10,788	4,649	4,293
1996	53,956	28,676	18,252	5,643	1,179	206	1996	50,581	12,386	18,217	11,084	4,731	4,163
1997	48,423	25,655	16,339	5,117	1,113	199	1997	53,999	13,155	19,714	11,928	4,875	4,327
1998	45,159	24,111	15,333	4,570	947	198	1998	56,821	14,254	21,136	12,431	4,918	4,082
1999	42,765	23,303	14,293	4,104	882	183	1999	58,932	15,571	22,133	12,439	4,939	3,850
2000	39,861	21,422	13,596	3,880	799	164	2000	61,758	16,352	23,355	13,121	4,959	3,971
2001	40,370	22,038	13,597	3,834	732	169	2001	63,353	16,747	24,650	13,120	4,931	3,905
2002	40,385	22,297	13,394	3,800	727	167	2002	66,161	18,088	25,842	13,170	5,009	4,052
2003	40,569	22,064	13,811	3,780	751	163	2003	70,886	19,811	27,569	14,185	5,096	4,225

1 See section 2.9.

Table 4.1   Live births within marriage: number of previous live-born children and age of mother (five-year age groups), 1993-2003
c. remarried women

England and Wales

Year	Number of previous live-born children[1]						Year	Number of previous live-born children[1]					
	Total	0	1	2	3	4 or more		Total	0	1	2	3	4 or more
All ages of mother at birth							**25-29**						
1993	**35,860**	9,517	11,602	8,202	4,022	2,517	1993	**8,710**	2,716	2,949	1,864	828	353
1994	**35,196**	9,580	11,488	7,781	3,886	2,461	1994	**8,079**	2,649	2,720	1,625	717	368
1995	**33,285**	9,155	10,836	7,397	3,615	2,282	1995	**7,211**	2,402	2,385	1,493	658	273
1996	**32,582**	9,026	10,557	7,197	3,492	2,310	1996	**6,395**	2,103	2,149	1,318	561	264
1997	**31,445**	8,630	10,249	6,795	3,492	2,279	1997	**5,828**	1,938	1,920	1,155	570	245
1998	**30,185**	8,501	9,983	6,594	3,035	2,072	1998	**5,105**	1,719	1,760	1,018	408	200
1999	**27,489**	7,923	9,056	5,731	2,872	1,907	1999	**4,254**	1,485	1,388	845	382	154
2000	**25,782**	7,299	8,574	5,469	2,591	1,849	2000	**3,719**	1,247	1,228	750	320	174
2001	**23,941**	6,933	8,066	4,966	2,305	1,671	2001	**3,135**	1,083	1,028	604	268	152
2002	**22,825**	6,751	7,717	4,547	2,237	1,573	2002	**2,682**	898	908	508	242	126
2003	**22,599**	6,708	7,597	4,656	2,163	1,475	2003	**2,402**	870	802	452	186	92
Under 20							**30-34**						
1993	**8**	2	2	1	2	1	1993	**14,821**	4,039	4,927	3,324	1,604	927
1994	**7**	5	2	-	-	-	1994	**14,920**	4,269	5,003	3,189	1,529	930
1995	**14**	8	4	2	-	-	1995	**14,030**	4,016	4,748	3,003	1,412	851
1996	**10**	5	3	2	-	-	1996	**13,925**	4,103	4,657	2,959	1,387	819
1997	**7**	5	2	-	-	-	1997	**13,068**	3,798	4,432	2,769	1,313	756
1998	**5**	2	2	-	1	-	1998	**12,389**	3,756	4,189	2,644	1,106	694
1999	**4**	3	1	-	-	-	1999	**11,324**	3,592	3,918	2,202	1,007	605
2000	**5**	4	1	-	-	-	2000	**10,354**	3,196	3,583	2,130	872	573
2001	**2**	1	1	-	-	-	2001	**9,460**	3,051	3,257	1,865	787	500
2002	**8**	4	2	2	-	-	2002	**8,849**	2,947	3,099	1,618	742	443
2003	**7**	3	3	1	-	-	2003	**8,435**	2,808	2,913	1,610	676	428
20-24							**35 and over**						
1993	**1,200**	405	418	271	74	32	1993	**11,121**	2,355	3,306	2,742	1,514	1,204
1994	**965**	304	370	218	57	16	1994	**11,225**	2,353	3,393	2,749	1,583	1,147
1995	**839**	284	303	175	58	19	1995	**11,191**	2,445	3,396	2,724	1,487	1,139
1996	**695**	249	229	146	52	19	1996	**11,557**	2,566	3,519	2,772	1,492	1,208
1997	**645**	236	219	140	40	10	1997	**11,897**	2,653	3,676	2,731	1,569	1,268
1998	**565**	192	204	109	46	14	1998	**12,121**	2,832	3,828	2,823	1,474	1,164
1999	**425**	171	150	71	28	5	1999	**11,482**	2,672	3,599	2,613	1,455	1,143
2000	**401**	149	148	70	28	6	2000	**11,303**	2,703	3,614	2,519	1,371	1,096
2001	**366**	150	126	63	19	8	2001	**10,978**	2,648	3,654	2,434	1,231	1,011
2002	**327**	135	112	66	12	2	2002	**10,959**	2,767	3,596	2,353	1,241	1,002
2003	**318**	123	104	54	26	11	2003	**11,437**	2,904	3,775	2,539	1,275	944

1 See section 2.9.

Table 4.2 Live births within marriage: number of previous live-born children and age of mother (single years), 2003 — **England and Wales**
a. all married women

Age of mother at birth	Number of previous live-born children[1]												
	Total	0	1	2	3	4	5	6	7	8	9	10-14[2]	15 and[2] over
All ages	364,244	150,982	132,873	51,981	17,641	6,155	2,534	1,046	503	274	121	133	1
Under 16	13	10	1	2	-	-	-	-	-	-	-	-	-
16	97	82	13	-	2	-	-	-	-	-	-	-	-
17	409	371	36	2	-	-	-	-	-	-	-	-	-
18	1,149	955	181	11	1	-	1	-	-	-	-	-	-
19	2,670	2,053	560	52	5	-	-	-	-	-	-	-	-
Under 20	4,338	3,471	791	67	8	-	1	-	-	-	-	-	-
20	4,304	3,029	1,099	146	25	2	1	1	-	-	-	1	-
21	6,227	3,743	2,042	384	42	12	3	1	-	-	-	-	-
22	8,014	4,277	2,882	700	139	11	5	-	-	-	-	-	-
23	10,423	5,263	3,738	1,157	210	48	6	1	-	-	-	-	-
24	11,919	5,875	4,154	1,447	361	66	14	2	-	-	-	-	-
20-24	40,887	22,187	13,915	3,834	777	139	29	5	-	-	-	1	-
25	13,972	6,901	4,616	1,788	536	104	22	5	-	-	-	-	-
26	16,536	8,240	5,273	2,130	666	163	47	14	2	1	-	-	-
27	19,443	9,631	6,293	2,419	801	230	46	18	3	-	-	2	-
28	22,766	11,126	7,480	2,754	981	288	101	29	4	2	-	1	-
29	25,977	12,541	8,818	3,017	1,088	365	115	25	6	1	-	1	-
25-29	98,694	48,439	32,480	12,108	4,072	1,150	331	91	15	4	-	4	-
30	28,088	12,859	10,117	3,321	1,150	429	138	52	17	2	1	2	-
31	29,603	12,591	11,330	3,680	1,273	446	201	56	24	2	-	-	-
32	29,526	11,371	12,019	4,015	1,354	483	171	64	27	16	4	2	-
33	26,713	9,464	10,898	4,163	1,324	505	218	73	38	19	7	4	-
34	24,072	7,885	9,979	4,069	1,312	498	175	93	37	17	2	5	-
30-34	138,002	54,170	54,343	19,248	6,413	2,361	903	338	143	56	14	13	-
35	20,670	6,407	8,273	3,874	1,249	477	220	83	50	24	10	3	-
36	17,057	4,843	6,774	3,457	1,146	444	212	82	56	19	13	11	-
37	13,653	3,658	5,237	2,947	1,048	379	194	93	46	28	13	10	-
38	10,567	2,674	4,033	2,239	929	328	161	85	47	42	12	17	-
39	7,648	2,020	2,802	1,594	696	237	144	68	38	28	9	12	-
35-39	69,595	19,602	27,119	14,111	5,068	1,865	931	411	237	141	57	53	-
40	5,192	1,296	1,856	1,091	486	200	116	54	39	24	14	15	1
41	3,289	774	1,099	692	327	177	90	61	24	14	12	19	-
42	1,959	459	634	394	237	111	47	27	20	12	12	6	-
43	1,107	233	331	244	139	65	38	23	13	11	4	6	-
44	566	137	165	111	57	40	17	17	9	4	3	6	-
40-44	12,113	2,899	4,085	2,532	1,246	593	308	182	105	65	45	52	1
45	272	98	63	43	22	16	8	11	2	4	3	2	-
46	136	42	32	16	15	14	9	-	1	2	1	4	-
47	71	21	13	7	13	8	3	3	-	-	-	3	-
48	42	16	11	2	1	4	2	3	-	1	1	1	-
49 and over	94	37	21	13	6	5	9	2	-	1	-	-	-
45 and over	615	214	140	81	57	47	31	19	3	8	5	10	-

Note: 1,874 cases in which age of mother at birth, and 84 cases in which the number of previous live-born children was not stated have been included with the stated cases (for method of distribution - see sections 3.2 and 3.3).
1 See section 2.9.
2 Detailed distribution for these groups is as follows:

Number of previous live-born children	Number of live births
10	64
11	35
12	17
13	13
14	4
15	1

Table 4.2 Live births within marriage: number of previous live-born children and age of mother (single years), 2003
b. women married once only

England and Wales

Age of mother at birth	Number of previous live-born children[1]												
	Total	0	1	2	3	4	5	6	7	8	9	10-14[2]	15 and[2] over
All ages	341,645	144,274	125,276	47,325	15,478	5,276	2,212	885	447	239	108	124	1
Under 16	13	10	1	2	-	-	-	-	-	-	-	-	-
16	97	82	13	-	2	-	-	-	-	-	-	-	-
17	409	371	36	2	-	-	-	-	-	-	-	-	-
18	1,145	952	181	10	1	-	1	-	-	-	-	-	-
19	2,667	2,053	557	52	5	-	-	-	-	-	-	-	-
Under 20	4,331	3,468	788	66	8	-	1	-	-	-	-	-	-
20	4,290	3,022	1,092	146	25	2	1	1	-	-	-	1	-
21	6,198	3,730	2,033	380	39	12	3	1	-	-	-	-	-
22	7,963	4,247	2,870	693	137	11	5	-	-	-	-	-	-
23	10,339	5,235	3,706	1,143	205	44	5	1	-	-	-	-	-
24	11,779	5,830	4,110	1,418	345	63	12	1	-	-	-	-	-
20-24	40,569	22,064	13,811	3,780	751	132	26	4	-	-	-	1	-
25	13,753	6,822	4,535	1,747	526	96	22	5	-	-	-	-	-
26	16,227	8,127	5,169	2,076	637	157	46	13	1	1	-	-	-
27	18,982	9,468	6,135	2,330	771	213	43	18	2	-	-	2	-
28	22,192	10,914	7,300	2,648	925	273	99	26	4	2	-	1	-
29	25,138	12,238	8,539	2,855	1,027	344	103	24	6	1	-	1	-
25-29	96,292	47,569	31,678	11,656	3,886	1,083	313	86	13	4	-	4	-
30	26,879	12,427	9,731	3,091	1,050	393	120	46	16	2	1	2	-
31	28,142	12,079	10,808	3,425	1,160	403	192	50	23	2	-	-	-
32	27,797	10,793	11,413	3,697	1,213	423	155	58	25	15	3	2	-
33	24,742	8,811	10,228	3,760	1,183	440	192	62	37	19	7	3	-
34	22,007	7,252	9,250	3,665	1,131	424	150	78	36	16	1	4	-
30-34	129,567	51,362	51,430	17,638	5,737	2,083	809	294	137	54	12	11	-
35	18,577	5,823	7,546	3,417	1,047	405	186	75	47	21	7	3	-
36	15,054	4,267	6,095	3,031	938	372	192	72	49	14	13	11	-
37	11,819	3,207	4,587	2,513	880	295	171	79	38	26	13	10	-
38	8,954	2,309	3,495	1,876	718	256	129	68	41	38	10	14	-
39	6,359	1,698	2,369	1,322	538	178	123	53	35	22	9	12	-
35-39	60,763	17,304	24,092	12,159	4,121	1,506	801	347	210	121	52	50	-
40	4,216	1,044	1,564	866	374	154	94	40	32	20	13	14	1
41	2,644	630	901	544	250	138	74	46	20	13	9	19	-
42	1,486	365	501	294	170	70	29	18	14	8	11	6	-
43	864	186	272	182	99	45	31	19	10	11	3	6	-
44	432	107	130	81	40	29	10	16	8	3	3	5	-
40-44	9,642	2,332	3,368	1,967	933	436	238	139	84	55	39	50	1
45	220	82	51	32	18	13	7	8	2	2	3	2	-
46	98	31	22	11	12	10	7	-	1	1	1	2	-
47	53	17	10	6	7	6	1	3	-	-	-	3	-
48	34	15	9	-	1	2	2	2	-	1	1	1	-
49 and over	76	30	17	10	4	5	7	2	-	1	-	-	-
45 and over	481	175	109	59	42	36	24	15	3	5	5	8	-

Note: 1,855 cases in which age of mother at birth, and 84 cases in which the number of previous live-born children was not stated have been included with the stated cases (for method of distribution - see sections 3.2 and 3.3).
1 See section 2.9.
2 Detailed distribution for these groups is as follows:

Number of previous live-born children	Number of live births
10	59
11	34
12	14
13	13
14	4
15	1

Table 4.2 Live births within marriage: number of previous live-born children and age of mother (single years), 2003
c. remarried women

Age of mother at birth	Number of previous live-born children[1]												
	Total	0	1	2	3	4	5	6	7	8	9	10-14[2]	15 and[2] over
All ages	22,599	6,708	7,597	4,656	2,163	879	322	161	56	35	13	9	-
Under 16	-	-	-	-	-	-	-	-	-	-	-	-	-
16	-	-	-	-	-	-	-	-	-	-	-	-	-
17	-	-	-	-	-	-	-	-	-	-	-	-	-
18	4	3	-	1	-	-	-	-	-	-	-	-	-
19	3	-	3	-	-	-	-	-	-	-	-	-	-
Under 20	7	3	3	1	-	-	-	-	-	-	-	-	-
20	14	7	7	-	-	-	-	-	-	-	-	-	-
21	29	13	9	4	3	-	-	-	-	-	-	-	-
22	51	30	12	7	2	-	-	-	-	-	-	-	-
23	84	28	32	14	5	4	1	-	-	-	-	-	-
24	140	45	44	29	16	3	2	1	-	-	-	-	-
20-24	318	123	104	54	26	7	3	1	-	-	-	-	-
25	219	79	81	41	10	8	-	-	-	-	-	-	-
26	309	113	104	54	29	6	1	1	1	-	-	-	-
27	461	163	158	89	30	17	3	-	1	-	-	-	-
28	574	212	180	106	56	15	2	3	-	-	-	-	-
29	839	303	279	162	61	21	12	1	-	-	-	-	-
25-29	2,402	870	802	452	186	67	18	5	2	-	-	-	-
30	1,209	432	386	230	100	36	18	6	1	-	-	-	-
31	1,461	512	522	255	113	43	9	6	1	-	-	-	-
32	1,729	578	606	318	141	60	16	6	2	1	1	-	-
33	1,971	653	670	403	141	65	26	11	1	-	-	1	-
34	2,065	633	729	404	181	74	25	15	1	1	1	1	-
30-34	8,435	2,808	2,913	1,610	676	278	94	44	6	2	2	2	-
35	2,093	584	727	457	202	72	34	8	3	3	3	-	-
36	2,003	576	679	426	208	72	20	10	7	5	-	-	-
37	1,834	451	650	434	168	84	23	14	8	2	-	-	-
38	1,613	365	538	363	211	72	32	17	6	4	2	3	-
39	1,289	322	433	272	158	59	21	15	3	6	-	-	-
35-39	8,832	2,298	3,027	1,952	947	359	130	64	27	20	5	3	-
40	976	252	292	225	112	46	22	14	7	4	1	1	-
41	645	144	198	148	77	39	16	15	4	1	3	-	-
42	473	94	133	100	67	41	18	9	6	4	1	-	-
43	243	47	59	62	40	20	7	4	3	-	1	-	-
44	134	30	35	30	17	11	7	1	1	1	-	1	-
40-44	2,471	567	717	565	313	157	70	43	21	10	6	2	-
45	52	16	12	11	4	3	1	3	-	2	-	-	-
46	38	11	10	5	3	4	2	-	-	1	-	2	-
47	18	4	3	1	6	2	2	-	-	-	-	-	-
48	8	1	2	2	-	2	-	1	-	-	-	-	-
49 and over	18	7	4	3	2	-	2	-	-	-	-	-	-
45 and over	134	39	31	22	15	11	7	4	-	3	-	2	-

Note: 19 cases in which age of mother at birth was not stated have been included with the stated cases (for method of distribution - see sections 3.2 and 3.3).
1 See section 2.9.
2 Detailed distribution for these groups is as follows:

Number of previous live-born children	Number of live births
10	5
11	1
12	3

Table 4.3　Live births and stillbirths within marriage (numbers and sex ratios):　　　　　　　　　**England and Wales**
number of previous live-born children, age of mother and sex, 2003

Age of mother at birth	Sex	Number of previous live-born children[1]												
		Total	0	1	2	3	4	5	6	7	8	9	10-14[2]	15 and over[2]
		Numbers												
All ages	M	**187,771**	77,596	68,759	26,649	9,139	3,224	1,348	510	266	135	61	83	1
	F	**178,434**	74,095	64,802	25,646	8,627	2,994	1,215	552	244	142	64	53	-
Under 20	M	**2,181**	1,749	399	27	5	-	1	-	-	-	-	-	-
	F	**2,193**	1,746	403	40	4	-	-	-	-	-	-	-	-
20-24	M	**21,068**	11,449	7,185	1,954	394	68	15	1	-	-	-	2	-
	F	**20,013**	10,828	6,793	1,913	388	71	16	4	-	-	-	-	-
25-29	M	**50,748**	24,817	16,761	6,213	2,113	612	169	48	10	2	-	3	-
	F	**48,481**	23,843	15,907	5,971	1,991	550	165	46	5	2	-	1	-
30-34	M	**71,280**	27,941	28,225	9,803	3,341	1,234	464	151	76	29	7	9	-
	F	**67,437**	26,472	26,373	9,570	3,123	1,151	449	190	70	28	7	4	-
35-39	M	**35,981**	10,029	14,027	7,325	2,623	993	520	215	120	66	31	32	-
	F	**33,993**	9,683	13,228	6,848	2,472	893	421	203	119	76	28	22	-
40-44	M	**6,190**	1,500	2,088	1,282	637	293	163	85	56	31	22	32	1
	F	**6,021**	1,420	2,030	1,268	618	305	149	100	50	35	25	21	-
45 and over	M	**323**	111	74	45	26	24	16	10	4	7	1	5	-
	F	**296**	103	68	36	31	24	15	9	-	1	4	5	-
Ratio: male births inside marriage per 1,000 female births inside marriage														
All ages		**1,052**	1,047	1,061	1,039	1,059	1,077	1,109	924	1,090	951	953	1,566	-
Under 20		**995**	1,002	990	675	*1,250*	-	-	-	-	-	-	-	-
20-24		**1,053**	1,057	1,058	1,021	1,015	958	*938*	250	-	-	-	-	-
25-29		**1,047**	1,041	1,054	1,041	1,061	1,113	1,024	1,043	*2,000*	*1,000*	-	*3,000*	-
30-34		**1,057**	1,055	1,070	1,024	1,070	1,072	1,033	795	1,086	1,036	*1,000*	*2,250*	-
35-39		**1,058**	1,036	1,060	1,070	1,061	1,112	1,235	1,059	1,008	868	1,107	1,455	-
40-44		**1,028**	1,056	1,029	1,011	1,031	961	1,094	850	1,120	886	880	1,524	-
45 and over		**1,091**	1,078	1,088	1,250	839	1,000	*1,067*	*1,111*	-	*7,000*	*250*	*1,000*	-

Note: 1,995 cases in which age of mother at birth, and 634 cases in which the number of previous live-born children was not stated hav▯▯▯▯▯▯▯▯▯▯▯▯▯▯▯ the stated cases (for method of distribution - see sections 3.2 and 3.3).
1 See section 2.9.
2 Detailed distribution for these groups is as follows:

Number of previous live-born children	Number of live births
10	65
11	36
12	18
13	13
14	4
15	1

Table 5.1 First live births within marriage: duration of marriage and age of mother, 1993-2003
a. all married women

England and Wales

Year	All durations	Completed months		Completed years													
		0-7	8-11	0	1	2	3	4	5	6	7	8	9	10-14	15-19	20 and over

All ages of mother at birth

Year	All durations	0-7	8-11	0	1	2	3	4	5	6	7	8	9	10-14	15-19	20 and over
1993	178,141	19,816	16,205	36,021	41,431	30,084	21,823	15,042	10,550	7,040	4,733	3,332	2,263	4,727	870	90
1994	176,046	18,578	15,765	34,343	42,371	29,388	20,874	15,368	10,403	7,153	4,914	3,179	2,281	4,823	830	119
1995	168,118	17,332	15,611	32,943	40,542	28,659	19,683	13,947	10,115	6,788	4,763	3,098	2,262	4,411	797	110
1996	163,020	17,300	15,489	32,789	39,521	27,283	18,951	13,225	9,333	6,806	4,622	3,129	2,173	4,274	807	107
1997	157,049	16,760	15,199	31,959	38,650	26,225	18,075	12,479	8,680	6,243	4,632	3,043	2,181	4,044	728	110
1998	155,708	16,582	15,617	32,199	38,631	26,169	17,648	12,097	8,643	5,886	4,350	3,091	2,006	4,165	720	103
1999	153,423	16,193	15,346	31,539	38,723	25,803	17,267	11,989	8,271	5,797	4,004	2,964	2,223	4,085	670	88
2000	146,509	14,423	14,838	29,261	38,069	25,331	16,239	11,166	7,561	5,422	3,858	2,734	2,046	4,165	575	82
2001	143,908	13,565	14,887	28,452	37,897	25,564	16,428	10,706	7,179	4,961	3,653	2,571	1,866	3,959	585	87
2002	145,241	13,871	14,904	28,775	38,634	26,372	16,566	10,889	7,265	4,961	3,451	2,396	1,736	3,544	539	113
2003	150,982	14,489	15,990	30,479	39,795	26,841	17,546	11,518	7,579	4,995	3,516	2,420	1,783	3,791	621	98

Under 20

Year	All durations	0-7	8-11	0	1	2	3	4	5	6	7	8	9	10-14	15-19	20 and over
1993	5,196	2,157	1,179	3,336	1,496	278	60	4	3	1	-	-	-	-	-	-
1994	4,687	1,872	1,094	2,966	1,384	267	59	10	-	1	1	-	-	-	-	-
1995	4,308	1,605	1,128	2,733	1,228	282	56	7	-	1	1	-	-	-	-	-
1996	4,223	1,567	1,149	2,716	1,180	256	61	8	2	-	-	-	-	-	-	-
1997	4,121	1,505	1,031	2,536	1,237	272	71	4	1	-	-	-	-	-	-	-
1998	4,188	1,490	1,123	2,613	1,223	280	63	7	2	-	-	-	-	-	-	-
1999	4,290	1,455	996	2,451	1,430	333	69	6	1	-	-	-	-	-	-	-
2000	3,810	1,225	903	2,128	1,221	383	75	3	-	-	-	-	-	-	-	-
2001	3,781	1,086	998	2,084	1,253	358	78	6	2	-	-	-	-	-	-	-
2002	3,817	1,119	888	2,007	1,366	363	73	8	-	-	-	-	-	-	-	-
2003	3,471	1,033	902	1,935	1,149	307	72	7	-	1	-	-	-	-	-	-

20-24

Year	All durations	0-7	8-11	0	1	2	3	4	5	6	7	8	9	10-14	15-19	20 and over
1993	40,399	7,192	5,678	12,870	13,345	7,812	3,995	1,611	546	141	33	8	4	8	-	-
1994	36,421	6,339	5,117	11,456	12,375	6,956	3,468	1,491	473	153	23	9	1	16	-	-
1995	32,340	5,783	4,755	10,538	10,893	6,153	2,964	1,215	407	122	32	6	4	4	-	-
1996	28,925	5,408	4,469	9,877	9,780	5,227	2,510	1,033	347	107	34	4	1	5	-	-
1997	25,891	4,913	4,135	9,048	8,733	4,629	2,113	910	305	112	28	8	3	2	-	-
1998	24,303	4,494	3,885	8,379	8,270	4,386	1,973	832	299	112	35	8	-	9	-	-
1999	23,474	4,336	3,706	8,042	8,000	4,165	1,952	803	353	100	36	7	7	9	-	-
2000	21,571	3,896	3,415	7,311	7,507	4,016	1,665	681	263	78	28	15	-	7	-	-
2001	22,188	3,755	3,514	7,269	7,756	4,233	1,817	696	272	100	34	6	1	4	-	-
2002	22,432	3,620	3,566	7,186	7,987	4,236	1,819	726	292	143	32	8	2	1	-	-
2003	22,187	3,815	3,613	7,428	7,594	4,091	1,843	761	303	120	29	12	-	6	-	-

25-29

Year	All durations	0-7	8-11	0	1	2	3	4	5	6	7	8	9	10-14	15-19	20 and over
1993	77,612	6,161	5,772	11,933	16,998	14,639	11,954	8,810	6,126	3,586	1,920	923	412	263	3	-
1994	75,660	5,760	5,826	11,586	17,636	13,990	11,238	8,670	5,741	3,478	1,865	827	402	224	3	-
1995	71,036	5,391	5,850	11,241	17,299	13,629	10,082	7,518	5,234	3,071	1,709	758	309	186	-	-
1996	67,178	5,396	5,695	11,091	16,590	12,889	9,552	6,846	4,604	2,915	1,530	699	286	176	-	-
1997	63,139	5,280	5,620	10,900	16,230	12,182	8,905	6,052	4,055	2,405	1,397	649	236	127	1	-
1998	60,602	5,377	5,804	11,181	15,895	11,552	8,269	5,593	3,746	2,137	1,238	609	230	148	4	-
1999	57,385	5,152	5,611	10,763	15,467	11,093	7,596	5,229	3,298	1,961	1,073	561	226	117	1	-
2000	52,673	4,352	5,306	9,658	14,859	10,595	7,010	4,538	2,740	1,606	871	427	220	149	-	-
2001	48,849	4,092	5,084	9,176	13,945	10,028	6,437	3,997	2,401	1,384	778	375	188	136	4	-
2002	47,111	4,060	4,967	9,027	13,592	9,802	6,310	3,763	2,177	1,208	630	324	157	121	-	-
2003	48,439	4,079	5,212	9,291	14,149	9,935	6,567	3,856	2,239	1,145	635	318	168	134	2	-

30 and over

Year	All durations	0-7	8-11	0	1	2	3	4	5	6	7	8	9	10-14	15-19	20 and over
1993	54,934	4,306	3,576	7,882	9,592	7,355	5,814	4,617	3,875	3,312	2,780	2,401	1,847	4,456	867	90
1994	59,278	4,607	3,728	8,335	10,976	8,175	6,109	5,197	4,189	3,521	3,026	2,343	1,878	4,583	827	119
1995	60,434	4,553	3,878	8,431	11,122	8,595	6,581	5,207	4,474	3,594	3,021	2,334	1,947	4,221	797	110
1996	62,694	4,929	4,176	9,105	11,971	8,911	6,828	5,338	4,380	3,784	3,058	2,426	1,886	4,093	807	107
1997	63,898	5,062	4,413	9,475	12,450	9,142	6,986	5,513	4,319	3,726	3,207	2,386	1,942	3,915	727	110
1998	66,615	5,221	4,805	10,026	13,243	9,951	7,343	5,665	4,596	3,637	3,077	2,474	1,776	4,008	716	103
1999	68,274	5,250	5,033	10,283	13,826	10,212	7,650	5,951	4,619	3,736	2,895	2,396	1,990	3,959	669	88
2000	68,455	4,950	5,214	10,164	14,482	10,337	7,489	5,944	4,558	3,738	2,959	2,292	1,826	4,009	575	82
2001	69,090	4,632	5,291	9,923	14,943	10,945	8,096	6,007	4,504	3,477	2,841	2,190	1,677	3,819	581	87
2002	71,881	5,072	5,483	10,555	15,689	11,971	8,364	6,392	4,796	3,610	2,789	2,064	1,577	3,422	539	113
2003	76,885	5,562	6,263	11,825	16,903	12,508	9,064	6,894	5,037	3,729	2,852	2,090	1,615	3,651	619	98

Table 5.1 First live births within marriage: duration of marriage and age of mother, 1993-2003
b. women married once only

England and Wales

Year	All durations	Completed months		Completed years										10-14	15-19	20 and over
		0-7	8-11	0	1	2	3	4	5	6	7	8	9			
All ages of mother at birth																
1993	**168,624**	17,678	14,928	32,606	38,823	28,756	21,037	14,558	10,262	6,847	4,589	3,247	2,205	4,620	860	90
1994	**166,466**	16,476	14,500	30,976	39,643	28,004	20,138	14,867	10,105	6,983	4,755	3,105	2,237	4,714	820	119
1995	**158,963**	15,368	14,320	29,688	38,073	27,298	18,947	13,461	9,799	6,609	4,660	3,023	2,200	4,304	791	110
1996	**153,994**	15,351	14,225	29,576	36,982	26,052	18,207	12,779	9,040	6,616	4,491	3,059	2,124	4,166	797	105
1997	**148,419**	14,878	14,049	28,927	36,178	24,992	17,418	12,048	8,401	6,069	4,526	2,966	2,128	3,938	719	109
1998	**147,207**	14,745	14,365	29,110	36,324	24,909	16,981	11,691	8,390	5,715	4,223	3,023	1,964	4,059	716	102
1999	**145,500**	14,545	14,186	28,731	36,489	24,678	16,594	11,595	8,010	5,653	3,914	2,895	2,181	4,007	665	88
2000	**139,210**	12,980	13,791	26,771	35,999	24,278	15,601	10,786	7,318	5,268	3,766	2,655	2,012	4,105	570	81
2001	**136,975**	12,323	13,894	26,217	35,888	24,468	15,807	10,348	6,975	4,823	3,567	2,510	1,829	3,876	580	87
2002	**138,490**	12,540	13,993	26,533	36,704	25,349	15,992	10,521	7,042	4,826	3,356	2,333	1,696	3,494	534	110
2003	**144,274**	13,166	15,025	28,191	37,893	25,844	16,951	11,190	7,351	4,868	3,428	2,362	1,753	3,731	616	96
Under 20																
1993	**5,194**	2,155	1,179	3,334	1,496	278	60	4	3	1	-	-	-	-	-	-
1994	**4,682**	1,870	1,094	2,964	1,381	267	59	10	-	1	-	-	-	-	-	-
1995	**4,300**	1,600	1,128	2,728	1,225	282	56	7	-	1	1	-	-	-	-	-
1996	**4,218**	1,565	1,146	2,711	1,180	256	61	8	2	-	-	-	-	-	-	-
1997	**4,116**	1,502	1,029	2,531	1,237	272	71	4	1	-	-	-	-	-	-	-
1998	**4,186**	1,489	1,123	2,612	1,222	280	63	7	2	-	-	-	-	-	-	-
1999	**4,287**	1,454	994	2,448	1,430	333	69	6	1	-	-	-	-	-	-	-
2000	**3,806**	1,223	902	2,125	1,220	383	75	3	-	-	-	-	-	-	-	-
2001	**3,780**	1,086	997	2,083	1,253	358	78	6	2	-	-	-	-	-	-	-
2002	**3,813**	1,118	888	2,006	1,364	362	73	8	-	-	-	-	-	-	-	-
2003	**3,468**	1,031	901	1,932	1,149	307	72	7	-	1	-	-	-	-	-	-
20-24																
1993	**39,994**	7,008	5,594	12,602	13,245	7,792	3,982	1,609	545	141	33	8	4	8	-	-
1994	**36,117**	6,214	5,058	11,272	12,290	6,929	3,462	1,490	472	153	23	9	1	16	-	-
1995	**32,056**	5,672	4,682	10,354	10,820	6,134	2,958	1,214	406	122	32	6	6	4	-	-
1996	**28,676**	5,299	4,421	9,720	9,713	5,210	2,502	1,033	347	107	34	4	1	5	-	-
1997	**25,655**	4,824	4,085	8,909	8,662	4,613	2,107	907	304	112	28	8	3	2	-	-
1998	**24,111**	4,423	3,836	8,259	8,221	4,368	1,969	831	299	112	35	8	-	9	-	-
1999	**23,303**	4,269	3,660	7,929	7,956	4,158	1,946	802	353	100	36	7	7	9	-	-
2000	**21,422**	3,848	3,375	7,223	7,461	4,006	1,661	680	263	78	28	15	-	7	-	-
2001	**22,038**	3,704	3,474	7,178	7,720	4,212	1,816	695	272	100	34	6	1	4	-	-
2002	**22,297**	3,572	3,535	7,107	7,948	4,224	1,815	725	292	143	32	8	2	1	-	-
2003	**22,064**	3,784	3,578	7,362	7,553	4,081	1,837	761	303	120	29	12	-	6	-	-
25-29																
1993	**74,896**	5,419	5,318	10,737	16,117	14,286	11,797	8,739	6,095	3,572	1,914	920	411	263	3	-
1994	**73,011**	5,078	5,346	10,424	16,782	13,619	11,101	8,594	5,711	3,465	1,862	827	401	222	3	-
1995	**68,634**	4,744	5,414	10,158	16,525	13,309	9,970	7,447	5,208	3,060	1,707	757	308	185	-	-
1996	**65,075**	4,817	5,310	10,127	15,934	12,625	9,427	6,791	4,585	2,905	1,526	699	285	171	-	-
1997	**61,201**	4,754	5,280	10,034	15,572	11,940	8,809	6,012	4,036	2,395	1,393	648	236	125	1	-
1998	**58,883**	4,890	5,458	10,348	15,386	11,338	8,169	5,557	3,734	2,130	1,233	609	230	145	4	-
1999	**55,900**	4,734	5,289	10,023	15,027	10,914	7,530	5,196	3,278	1,956	1,071	561	226	117	1	-
2000	**51,426**	4,035	5,067	9,102	14,457	10,428	6,945	4,502	2,729	1,601	869	426	220	147	-	-
2001	**47,766**	3,824	4,866	8,690	13,589	9,892	6,373	3,974	2,391	1,379	775	375	188	136	4	-
2002	**46,213**	3,801	4,811	8,612	13,305	9,696	6,261	3,737	2,168	1,206	628	322	157	121	-	-
2003	**47,569**	3,838	5,033	8,871	13,905	9,821	6,512	3,831	2,232	1,143	632	318	168	134	2	-
30 and over																
1993	**48,540**	3,096	2,837	5,933	7,965	6,400	5,198	4,206	3,619	3,133	2,642	2,319	1,790	4,349	857	90
1994	**52,656**	3,314	3,002	6,316	9,190	7,189	5,516	4,773	3,922	3,364	2,870	2,269	1,835	4,476	817	119
1995	**53,973**	3,352	3,096	6,448	9,503	7,573	5,963	4,793	4,185	3,426	2,920	2,260	1,886	4,115	791	110
1996	**56,025**	3,670	3,348	7,018	10,155	7,961	6,217	4,947	4,106	3,604	2,931	2,356	1,838	3,990	797	105
1997	**57,447**	3,798	3,655	7,453	10,707	8,167	6,431	5,125	4,060	3,562	3,105	2,310	1,889	3,811	718	109
1998	**60,027**	3,943	3,948	7,891	11,495	8,923	6,780	5,296	4,355	3,473	2,955	2,406	1,734	3,905	712	102
1999	**62,010**	4,088	4,243	8,331	12,076	9,273	7,049	5,591	4,378	3,597	2,807	2,327	1,948	3,881	664	88
2000	**62,556**	3,874	4,447	8,321	12,861	9,461	6,920	5,601	4,326	3,589	2,869	2,214	1,792	3,951	570	81
2001	**63,391**	3,709	4,557	8,266	13,326	10,006	7,540	5,673	4,310	3,344	2,758	2,129	1,640	3,736	576	87
2002	**66,167**	4,049	4,759	8,808	14,087	11,067	7,843	6,051	4,582	3,477	2,696	2,003	1,537	3,372	534	110
2003	**71,173**	4,513	5,513	10,026	15,286	11,635	8,530	6,591	4,816	3,604	2,767	2,032	1,585	3,591	614	96

Table 5.1 First live births within marriage: duration of marriage and age of mother, 1993-2003
c. remarried women

England and Wales

Year	All durations	Completed months		Completed years												
		0-7	8-11	0	1	2	3	4	5	6	7	8	9	10-14	15-19	20 and over
All ages of mother at birth																
1993	**9,517**	2,138	1,277	3,415	2,608	1,328	786	484	288	193	144	85	58	107	10	-
1994	**9,580**	2,102	1,265	3,367	2,728	1,384	736	501	298	170	159	74	44	109	10	-
1995	**9,155**	1,964	1,291	3,255	2,469	1,361	736	486	316	179	103	75	62	107	6	-
1996	**9,026**	1,949	1,264	3,213	2,539	1,231	744	446	293	190	131	70	49	108	10	2
1997	**8,630**	1,882	1,150	3,032	2,472	1,233	657	431	279	174	106	77	53	106	9	1
1998	**8,501**	1,837	1,252	3,089	2,307	1,260	667	406	253	171	127	68	42	106	4	1
1999	**7,923**	1,648	1,160	2,808	2,234	1,125	673	394	261	144	90	69	42	78	5	-
2000	**7,299**	1,443	1,047	2,490	2,070	1,053	638	380	243	154	92	79	34	60	5	1
2001	**6,933**	1,242	993	2,235	2,009	1,096	621	358	204	138	86	61	37	83	5	-
2002	**6,751**	1,331	911	2,242	1,930	1,023	574	368	223	135	95	63	40	50	5	3
2003	**6,708**	1,323	965	2,288	1,902	997	595	328	228	127	88	58	30	60	5	2
Under 20																
1993	**2**	2	-	2	-	-	-	-	-	-	-	-	-	-	-	-
1994	**5**	2	-	2	3	-	-	-	-	-	-	-	-	-	-	-
1995	**8**	5	-	5	3	-	-	-	-	-	-	-	-	-	-	-
1996	**5**	2	3	5	-	-	-	-	-	-	-	-	-	-	-	-
1997	**5**	3	2	5	-	-	-	-	-	-	-	-	-	-	-	-
1998	**2**	1	-	1	1	-	-	-	-	-	-	-	-	-	-	-
1999	**3**	1	2	3	-	-	-	-	-	-	-	-	-	-	-	-
2000	**4**	2	1	3	1	-	-	-	-	-	-	-	-	-	-	-
2001	**1**	-	1	1	-	-	-	-	-	-	-	-	-	-	-	-
2002	**4**	1	-	1	2	1	-	-	-	-	-	-	-	-	-	-
2003	**3**	2	1	3	-	-	-	-	-	-	-	-	-	-	-	-
20-24																
1993	**405**	184	84	268	100	20	13	2	1	-	-	-	-	-	-	-
1994	**304**	125	59	184	85	27	6	1	1	-	-	-	-	-	-	-
1995	**284**	111	73	184	73	19	6	1	1	-	-	-	-	-	-	-
1996	**249**	109	48	157	67	17	8	-	-	-	-	-	-	-	-	-
1997	**236**	89	50	139	71	16	6	3	1	-	-	-	-	-	-	-
1998	**192**	71	49	120	49	18	4	1	-	-	-	-	-	-	-	-
1999	**171**	67	46	113	44	7	6	1	-	-	-	-	-	-	-	-
2000	**149**	48	40	88	46	10	4	1	-	-	-	-	-	-	-	-
2001	**150**	51	40	91	36	21	1	1	-	-	-	-	-	-	-	-
2002	**135**	48	31	79	39	12	4	1	-	-	-	-	-	-	-	-
2003	**123**	31	35	66	41	10	6	-	-	-	-	-	-	-	-	-
25-29																
1993	**2,716**	742	454	1,196	881	353	157	71	31	14	6	3	1	-	-	-
1994	**2,649**	682	480	1,162	854	371	137	76	30	13	3	-	1	2	-	-
1995	**2,402**	647	436	1,083	774	320	112	71	26	11	2	1	1	1	-	-
1996	**2,103**	579	385	964	656	264	125	55	19	10	4	-	1	5	-	-
1997	**1,938**	526	340	866	658	242	96	40	19	10	4	1	-	2	-	-
1998	**1,719**	487	346	833	509	214	100	36	12	7	5	-	-	3	-	-
1999	**1,485**	418	322	740	440	179	66	33	20	5	2	-	-	-	-	-
2000	**1,247**	317	239	556	402	167	65	36	11	5	2	1	-	2	-	-
2001	**1,083**	268	218	486	356	136	64	23	10	5	3	-	-	-	-	-
2002	**898**	259	156	415	287	106	49	26	9	2	2	2	-	-	-	-
2003	**870**	241	179	420	244	114	55	25	7	2	3	-	-	-	-	-
30 and over																
1993	**6,394**	1,210	739	1,949	1,627	955	616	411	256	179	138	82	57	107	10	-
1994	**6,622**	1,293	726	2,019	1,786	986	593	424	267	157	156	74	43	107	10	-
1995	**6,461**	1,201	782	1,983	1,619	1,022	618	414	289	168	101	74	61	106	6	-
1996	**6,669**	1,259	828	2,087	1,816	950	611	391	274	180	127	70	48	103	10	2
1997	**6,451**	1,264	758	2,022	1,743	975	555	388	259	164	102	76	53	104	9	1
1998	**6,588**	1,278	857	2,135	1,748	1,028	563	369	241	164	122	68	42	103	4	1
1999	**6,264**	1,162	790	1,952	1,750	939	601	360	241	139	88	69	42	78	5	-
2000	**5,899**	1,076	767	1,843	1,621	876	569	343	232	149	90	78	34	58	5	1
2001	**5,699**	923	734	1,657	1,617	939	556	334	194	133	83	61	37	83	5	-
2002	**5,714**	1,023	724	1,747	1,602	904	521	341	214	133	93	61	40	50	5	3
2003	**5,712**	1,049	750	1,799	1,617	873	534	303	221	125	85	58	30	60	5	2

Table 5.2 Live births within 8 months of marriage: order of marriage and age of mother, 1993-2003　　　　　　　　　　**England and Wales**

Age of mother at birth	1993	1994	1995	1996	1997	1998	1999	2000	2001	2002	2003
All married women											
All ages	**28,249**	**27,242**	**25,454**	**25,265**	**24,467**	**24,042**	**23,196**	**20,806**	**19,679**	**19,745**	**20,561**
Under 16	4	3	2	1	4	1	1	-	-	4	1
16	96	78	61	86	73	48	54	42	57	48	37
17	308	269	250	225	221	223	201	156	156	157	134
18	788	731	587	546	595	569	552	467	401	429	376
19	1,291	1,052	960	925	818	861	842	737	618	611	639
Under 20	**2,487**	**2,133**	**1,860**	**1,783**	**1,711**	**1,702**	**1,650**	**1,402**	**1,232**	**1,249**	**1,187**
20	1,588	1,413	1,238	1,139	1,010	913	988	904	868	805	763
21	1,915	1,660	1,416	1,306	1,167	1,145	1,059	1,033	987	920	1,025
22	1,984	1,793	1,590	1,448	1,274	1,202	1,080	975	1,041	1,042	957
23	1,964	1,918	1,681	1,608	1,408	1,296	1,204	1,043	1,063	1,028	1,179
24	2,016	1,786	1,762	1,684	1,603	1,453	1,353	1,163	1,031	1,024	1,059
20-24	**9,467**	**8,570**	**7,687**	**7,185**	**6,462**	**6,009**	**5,684**	**5,118**	**4,990**	**4,819**	**4,983**
25-29	8,967	8,622	7,979	8,035	7,708	7,612	7,294	6,266	5,838	5,609	5,692
30-34	4,961	5,498	5,370	5,531	5,766	5,656	5,541	5,087	4,824	5,007	5,163
35-39	1,922	2,029	2,121	2,264	2,310	2,546	2,501	2,432	2,270	2,479	2,858
40 and over	445	390	437	467	510	517	526	501	525	582	678
Women married once only											
All ages	**22,673**	**21,683**	**20,250**	**20,175**	**19,596**	**19,438**	**19,169**	**17,210**	**16,530**	**16,613**	**17,389**
Under 16	4	3	2	1	4	1	1	-	-	3	1
16	96	78	61	86	73	48	54	42	57	48	37
17	308	269	250	225	220	222	201	156	156	157	134
18	786	731	584	546	595	568	552	467	401	429	374
19	1,288	1,049	956	919	815	861	841	735	618	610	637
Under 20	**2,482**	**2,130**	**1,853**	**1,777**	**1,707**	**1,700**	**1,649**	**1,400**	**1,232**	**1,247**	**1,183**
20	1,574	1,405	1,232	1,134	1,003	906	980	899	867	798	757
21	1,884	1,621	1,395	1,280	1,146	1,130	1,045	1,024	979	906	1,015
22	1,911	1,733	1,548	1,406	1,251	1,177	1,061	948	1,021	1,031	945
23	1,851	1,820	1,582	1,523	1,357	1,252	1,162	1,011	1,030	1,002	1,152
24	1,812	1,647	1,628	1,570	1,500	1,367	1,285	1,099	982	981	1,031
20-24	**9,032**	**8,226**	**7,385**	**6,913**	**6,257**	**5,832**	**5,533**	**4,981**	**4,879**	**4,718**	**4,900**
25-29	7,140	6,877	6,440	6,604	6,461	6,511	6,370	5,516	5,223	5,025	5,168
30-34	2,977	3,371	3,362	3,538	3,796	3,856	3,935	3,634	3,579	3,802	3,970
35-39	858	944	1,023	1,143	1,158	1,344	1,443	1,434	1,352	1,527	1,801
40 and over	184	135	187	200	217	195	239	245	265	294	367
Remarried women											
All ages	**5,576**	**5,559**	**5,204**	**5,090**	**4,871**	**4,604**	**4,027**	**3,596**	**3,149**	**3,132**	**3,172**
Under 16	-	-	-	-	-	-	-	-	-	1	-
16	-	-	-	-	-	-	-	-	-	-	-
17	-	-	-	-	1	1	-	-	-	-	-
18	2	-	3	-	-	1	-	-	-	-	2
19	3	3	4	6	3	-	1	2	-	1	2
Under 20	**5**	**3**	**7**	**6**	**4**	**2**	**1**	**2**	**-**	**2**	**4**
20	14	8	6	5	7	7	8	5	1	7	6
21	31	39	21	26	21	15	14	9	8	14	10
22	73	60	42	42	23	25	19	27	20	11	12
23	113	98	99	85	51	44	42	32	33	26	27
24	204	139	134	114	103	86	68	64	49	43	28
20-24	**435**	**344**	**302**	**272**	**205**	**177**	**151**	**137**	**111**	**101**	**83**
25-29	1,827	1,745	1,539	1,431	1,247	1,101	924	750	615	584	524
30-34	1,984	2,127	2,008	1,993	1,970	1,800	1,606	1,453	1,245	1,205	1,193
35-39	1,064	1,085	1,098	1,121	1,152	1,202	1,058	998	918	952	1,057
40 and over	261	255	250	267	293	322	287	256	260	288	311

Table 5.3 Live births within 8 months of marriage: duration of marriage, order of marriage and age of mother, 2003 — England and Wales

Age of mother at birth	Duration of current marriage - completed months								
	0-7	0	1	2	3	4	5	6	7
All married women									
All ages	**20,561**	**950**	**1,432**	**1,900**	**2,648**	**3,379**	**3,792**	**3,378**	**3,082**
Under 16	1	-	-	-	-	-	1	-	-
16	37	4	5	7	5	4	9	1	2
17	134	11	27	7	17	20	18	21	13
18	376	27	44	53	59	52	56	45	40
19	639	30	60	66	87	102	116	91	87
Under 20	**1,187**	**72**	**136**	**133**	**168**	**178**	**200**	**158**	**142**
20	763	45	66	78	99	134	118	121	102
21	1,025	51	79	110	136	175	197	140	137
22	957	60	71	113	136	149	167	126	135
23	1,179	41	86	113	146	182	260	179	172
24	1,059	38	57	99	117	188	189	191	180
20-24	**4,983**	**235**	**359**	**513**	**634**	**828**	**931**	**757**	**726**
25-29	5,692	235	372	469	669	912	1,066	1,042	927
30-34	5,163	217	301	425	705	875	940	855	845
35-39	2,858	163	202	274	363	466	546	477	367
40 and over	678	28	62	86	109	120	109	89	75
Women married once only									
All ages	**17,389**	**812**	**1,197**	**1,593**	**2,206**	**2,829**	**3,243**	**2,864**	**2,645**
Under 16	1	-	-	-	-	-	1	-	-
16	37	4	5	7	5	4	9	1	2
17	134	11	27	7	17	20	18	21	13
18	374	27	44	53	59	52	56	44	39
19	637	30	60	65	87	102	116	91	86
Under 20	**1,183**	**72**	**136**	**132**	**168**	**178**	**200**	**157**	**140**
20	757	45	64	78	99	133	117	120	101
21	1,015	51	79	107	134	175	196	139	134
22	945	59	68	111	134	148	166	125	134
23	1,152	39	84	109	141	179	256	175	169
24	1,031	37	56	94	113	182	187	184	178
20-24	**4,900**	**231**	**351**	**499**	**621**	**817**	**922**	**743**	**716**
25-29	5,168	200	332	410	595	830	986	965	850
30-34	3,970	169	217	322	542	667	724	661	668
35-39	1,801	122	128	186	221	269	350	287	238
40 and over	367	18	33	44	59	68	61	51	33
Remarried women									
All ages	**3,172**	**138**	**235**	**307**	**442**	**550**	**549**	**514**	**437**
Under 16	-	-	-	-	-	-	-	-	-
16	-	-	-	-	-	-	-	-	-
17	-	-	-	-	-	-	-	-	-
18	2	-	-	-	-	-	-	1	1
19	2	-	-	1	-	-	-	-	1
Under 20	**4**	**-**	**-**	**1**	**-**	**-**	**-**	**1**	**2**
20	6	-	2	-	-	1	1	1	1
21	10	-	-	3	2	-	1	1	3
22	12	1	3	2	2	1	1	1	1
23	27	2	2	4	5	3	4	4	3
24	28	1	1	5	4	6	2	7	2
20-24	**83**	**4**	**8**	**14**	**13**	**11**	**9**	**14**	**10**
25-29	524	35	40	59	74	82	80	77	77
30-34	1,193	48	84	103	163	208	216	194	177
35-39	1,057	41	74	88	142	197	196	190	129
40 and over	311	10	29	42	50	52	48	38	42

Table 6.1 **Maternities with multiple births: occurrence within/outside marriage and age of mother, 1993-2003**
a. numbers

England and Wales

Year	Age of mother at birth							
	All ages	Under 20	20-24	25-29	30-34	35-39	40-44	45 and over
All maternities with multiple births								
1993	**8,549**	286	1,407	2,800	2,794	1,096	154	12
1994	**8,719**	277	1,269	2,836	2,914	1,263	150	10
1995	**9,038**	291	1,306	2,906	2,993	1,348	175	19
1996	**8,883**	273	1,152	2,643	3,192	1,408	194	21
1997	**9,217**	263	1,124	2,615	3,360	1,599	233	23
1998	**9,080**	303	1,078	2,457	3,332	1,627	252	31
1999	**8,907**	291	958	2,412	3,250	1,728	239	29
2000	**8,792**	297	962	2,163	3,246	1,835	264	25
2001	**8,700**	278	1,003	2,054	3,256	1,785	285	39
2002	**8,861**	301	1,095	1,987	3,165	1,952	314	47
2003	**9,131**	294	1,071	2,006	3,282	2,068	348	62
Maternities within marriage with multiple births								
1993	**6,209**	39	729	2,123	2,284	909	115	10
1994	**6,271**	34	656	2,094	2,362	1,003	112	10
1995	**6,498**	50	623	2,146	2,437	1,100	126	16
1996	**6,323**	28	501	1,914	2,590	1,120	152	18
1997	**6,369**	31	458	1,813	2,653	1,227	174	13
1998	**6,227**	41	427	1,668	2,606	1,269	192	24
1999	**6,011**	41	397	1,593	2,466	1,322	169	23
2000	**5,980**	35	347	1,432	2,539	1,408	195	24
2001	**5,837**	37	375	1,335	2,486	1,376	199	29
2002	**5,929**	26	431	1,327	2,437	1,453	220	35
2003	**5,987**	28	390	1,278	2,498	1,487	261	45
Maternities outside marriage with multiple births								
1993	**2,340**	247	678	677	510	187	39	2
1994	**2,448**	243	613	742	552	260	38	-
1995	**2,540**	241	683	760	556	248	49	3
1996	**2,560**	245	651	729	602	288	42	3
1997	**2,848**	232	666	802	707	372	59	10
1998	**2,853**	262	651	789	726	358	60	7
1999	**2,896**	250	561	819	784	406	70	6
2000	**2,812**	262	615	731	707	427	69	1
2001	**2,863**	241	628	719	770	409	86	10
2002	**2,932**	275	664	660	728	499	94	12
2003	**3,144**	266	681	728	784	581	87	17

Note: The figures include maternities where live births and/or stillbirths occurred.

Table 6.1   Maternities with multiple births: occurrence
within/outside marriage and age of mother, 1993-2003
b. rates[1]

England and Wales

Year	Age of mother at birth							
	All ages	Under 20	20-24	25-29	30-34	35-39	40-44	45 and over
	All maternities with multiple births per 1,000 all maternities							
1993	**12.8**	6.3	9.3	12.0	16.5	18.9	15.5	22.3
1994	**13.2**	6.6	9.1	12.5	16.4	20.3	14.7	20.7
1995	**14.1**	6.9	10.0	13.5	16.7	20.9	16.3	36.5
1996	**13.8**	6.1	9.2	12.6	17.4	20.6	17.0	36.9
1997	**14.5**	5.7	9.5	13.0	18.2	21.7	19.1	40.6
1998	**14.4**	6.3	9.5	12.8	17.9	21.0	19.7	56.7
1999	**14.5**	6.0	8.7	13.4	17.8	21.6	17.7	47.4
2000	**14.7**	6.5	9.0	12.8	18.3	22.0	18.5	38.8
2001	**14.8**	6.3	9.3	12.9	18.5	21.0	18.6	53.1
2002	**15.0**	6.9	9.9	13.1	17.8	21.9	19.3	55.3
2003	**14.8**	6.6	9.2	12.9	17.7	21.6	19.3	75.7
	Maternities within marriage with multiple births per 1,000 maternities within marriage							
1993	**13.7**	5.7	9.5	12.0	16.5	19.6	15.2	23.4
1994	**14.1**	5.6	9.5	12.4	16.4	20.5	14.7	27.9
1995	**15.3**	8.9	10.3	13.7	17.1	21.9	15.9	40.9
1996	**15.3**	5.2	9.2	13.0	18.0	21.4	18.2	41.3
1997	**15.9**	5.9	9.4	13.1	18.5	22.0	20.0	31.0
1998	**15.9**	7.8	9.4	12.9	18.3	21.8	21.2	58.4
1999	**16.0**	7.7	9.2	13.3	17.8	22.2	18.0	49.9
2000	**16.5**	7.4	8.6	12.9	18.9	22.9	19.9	52.2
2001	**16.6**	8.0	9.2	13.1	18.9	22.2	19.0	54.6
2002	**16.9**	5.7	10.6	13.7	18.4	22.6	20.0	57.7
2003	**16.6**	6.4	9.6	13.1	18.3	21.7	21.8	78.5
	Maternities outside marriage with multiple births per 1,000 maternities outside marriage							
1993	**10.9**	6.5	9.1	11.8	16.4	15.8	16.6	18.0
1994	**11.4**	6.8	8.6	12.8	16.4	19.6	14.8	-
1995	**11.6**	6.6	9.8	12.9	15.2	17.4	17.4	23.3
1996	**11.1**	6.2	9.2	11.8	15.0	17.9	13.6	22.6
1997	**12.0**	5.6	9.6	12.7	16.9	20.7	16.8	68.5
1998	**11.9**	6.1	9.6	12.7	16.7	18.5	16.0	51.5
1999	**12.0**	5.8	8.3	13.5	17.6	19.8	17.0	39.7
2000	**11.9**	6.4	9.2	12.5	16.3	19.4	15.5	5.4
2001	**12.1**	6.1	9.3	12.8	17.2	17.8	17.7	49.0
2002	**12.2**	7.1	9.5	11.9	15.8	20.2	17.8	49.4
2003	**12.3**	6.7	9.0	12.6	16.1	21.2	14.4	69.1

Note: The figures include maternities where live births and/or stillbirths occurred.
1 See section 2.11.

Table 6.2   Maternities with multiple births: whether live births   **England and Wales**
              or stillbirths, multiplicity, age of mother and sex, 2003

Age of mother at birth	Maternities	Births					
		Live			Still		
		Total	Male	Female	Total	Male	Female
All maternities with multiple births[1]							
All ages	**9,131**	**18,059**	**9,084**	**8,975**	**339**	**188**	**151**
Under 20	294	**573**	269	304	**20**	6	14
20-24	1,071	**2,118**	1,042	1,076	**37**	22	15
25-29	2,006	**3,957**	2,005	1,952	**82**	44	38
30-34	3,282	**6,489**	3,247	3,242	**120**	68	52
35-39	2,068	**4,108**	2,093	2,015	**66**	39	27
40-44	348	**691**	360	331	**12**	8	4
45 and over	62	**123**	68	55	**2**	1	1
Twins only							
All ages	**9,001**	**17,689**	**8,904**	**8,785**	**316**	**172**	**144**
Under 20	289	**558**	265	293	**20**	6	14
20-24	1,058	**2,079**	1,021	1,058	**37**	22	15
25-29	1,979	**3,878**	1,971	1,907	**80**	44	36
30-34	3,239	**6,371**	3,189	3,182	**108**	61	47
35-39	2,034	**4,013**	2,042	1,971	**57**	30	27
40-44	341	**670**	350	320	**12**	8	4
45 and over	61	**120**	66	54	**2**	1	1
Triplets only							
All ages	**127**	**361**	**176**	**185**	**20**	**13**	**7**
Under 20	5	**15**	4	11	-	-	-
20-24	13	**39**	21	18	-	-	-
25-29	27	**79**	34	45	**2**	-	2
30-34	42	**114**	57	57	**12**	7	5
35-39	32	**90**	48	42	**6**	6	-
40-44	7	**21**	10	11	-	-	-
45 and over	1	**3**	2	1	-	-	-

1 Includes quads and above.

Table 6.3   Maternities within marriage with multiple births (numbers and rates[1]):   **England and Wales**
              age of mother and number of previous live-born children, 2003

Number of previous live-born children	Age of mother at birth							
	All ages	Under 20	20-24	25-29	30-34	35-39	40-44	45 and over
Numbers								
Total	**5,987**	**28**	**390**	**1,278**	**2,498**	**1,487**	**261**	**45**
0	**2,892**	18	213	697	1,243	599	96	26
1	**1,906**	9	118	348	824	512	83	12
2	**765**	1	45	154	278	236	49	2
3	**267**	-	8	59	94	85	20	1
4 and over	**157**	-	6	20	59	55	13	4
Rates: maternities with multiple births per 1,000 maternities within marriage								
Total	**16.6**	**6.4**	**9.6**	**13.1**	**18.3**	**21.7**	**21.8**	**78.5**
0	**19.4**	*5.2*	9.7	14.5	23.4	31.4	34.0	139.0
1	**14.5**	*11.3*	8.5	10.8	15.3	19.2	20.6	*92.3*
2	**14.8**	*15.2*	11.8	12.8	14.6	16.9	19.6	*25.3*
3	**15.3**	-	*10.3*	14.6	14.8	17.0	16.2	*17.9*
4 and over	**14.6**	-	*35.1*	12.6	15.5	14.9	9.6	*33.1*

Note: The figures include maternities where live births and/or stillbirths occurred.
1 See section 2.11.

Table 6.4 All maternities: age of mother, multiplicity and type of outcome, 2003   **England and Wales**

Outcome	All maternities							
	All ages	Under 20	20-24	25-29	30-34	35-39	40-44	45 and over
All maternities	**615,787**	**44,245**	**116,147**	**155,802**	**184,935**	**95,841**	**17,998**	**819**
Singleton maternities	**606,656**	**43,951**	**115,076**	**153,796**	**181,653**	**93,773**	**17,650**	**757**
1 LM	**309,344**	22,478	58,800	78,206	92,763	47,852	8,875	370
1 LF	**294,066**	21,185	55,704	74,768	87,962	45,426	8,639	382
1 SM	**1,686**	154	304	423	491	240	70	4
1 SF	**1,560**	134	268	399	437	255	66	1
All multiple maternities	**9,131**	**294**	**1,071**	**2,006**	**3,282**	**2,068**	**348**	**62**
Twins	**9,001**	**289**	**1,058**	**1,979**	**3,239**	**2,034**	**341**	**61**
2 LM	**2,866**	88	346	678	1,010	617	109	18
1 LM and 1 LF	**3,033**	82	313	590	1,120	773	127	28
2 LF	**2,822**	101	367	643	1,015	589	94	13
1 LM and 1 SM	**110**	6	14	22	39	24	4	1
1 LM and 1 SF	**29**	1	2	3	10	11	1	1
1 LF and 1 SM	**33**	-	3	11	11	6	2	-
1 LF and 1 SF	**72**	9	8	20	20	12	3	-
2 SM	**12**	-	2	5	4	-	1	-
1 SM and 1 SF	**5**	-	1	1	3	-	-	-
2 SF	**19**	2	2	6	7	2	-	-
Triplets	**127**	**5**	**13**	**27**	**42**	**32**	**7**	**1**
3 LM	**23**	1	5	4	6	6	1	-
2 LM and 1 LF	**27**	-	2	6	9	8	1	1
1 LM and 2 LF	**36**	1	2	8	11	9	5	-
3 LF	**26**	3	4	7	7	5	-	-
2 LM and 1 SM	**3**	-	-	-	2	1	-	-
2 LM and 1 SF	**2**	-	-	1	1	-	-	-
1 LM, 1 LF and 1 SM	**2**	-	-	-	1	1	-	-
1 LM, 1 LF and 1 SF	**1**	-	-	-	1	-	-	-
2 LF and 1 SM	**-**	-	-	-	-	-	-	-
2 LF and 1 SF	**2**	-	-	1	1	-	-	-
1 LM and 2 SM	**3**	-	-	-	1	2	-	-
1 LM, 1 SM and 1 SF	**-**	-	-	-	-	-	-	-
1 LM and 2 SF	**1**	-	-	-	1	-	-	-
1 LF and 2 SM	**1**	-	-	-	1	-	-	-
1 LF, 1 SM and 1 SF	**-**	-	-	-	-	-	-	-
1 LF and 2 SF	**-**	-	-	-	-	-	-	-
3 SM	**-**	-	-	-	-	-	-	-
2 SM and 1 SF	**-**	-	-	-	-	-	-	-
1 SM and 2 SF	**-**	-	-	-	-	-	-	-
3 SF	**-**	-	-	-	-	-	-	-
Quads and above	**3**	**-**	**-**	**-**	**1**	**2**	**-**	**-**
3 LM and 1SM	**1**	-	-	-	-	1	-	-
1 LM and 3LF	**1**	-	-	-	1	-	-	-
2LF and 2SM	**1**	-	-	-	-	1	-	-

LM - Live-born male   SM - Stillborn male   LF - Live-born female   SF - Stillborn female

Table 7.1 Live births (numbers, rates, general fertility rate, and total fertility rate), and population: occurrence within/outside marriage, and area of usual residence, 2003

England and Wales, Government Office Regions (within England), counties and unitary authorities

Area of usual residence of mother	Estimated number of women aged 15-44 (000s)	Live births			Live births outside marriage per 1,000 live births	General fertility rate (GFR)[1]	Total fertility rate (TFR)[1]
		Total	Within marriage	Outside marriage			
ENGLAND AND WALES	**10,944.1**	**621,469**	**364,244**	**257,225**	**414**	**56.8**	**1.73**
ENGLAND	10,366.0	589,851	348,485	241,366	409	56.9	1.73
NORTH EAST	**519.0**	**27,005**	**12,554**	**14,451**	**535**	**52.0**	**1.66**
Darlington UA	19.7	**1,179**	586	593	503	59.9	1.95
Hartlepool UA	18.4	**1,065**	406	659	619	57.9	1.93
Middlesbrough UA	29.9	**1,786**	743	1,043	584	59.7	1.89
Redcar and Cleveland UA	27.3	**1,446**	570	876	606	53.0	1.77
Stockton-on-Tees UA	39.3	**2,115**	1,077	1,038	491	53.8	1.73
Durham	99.0	**4,929**	2,274	2,655	539	49.8	1.60
Northumberland	57.0	**2,934**	1,585	1,349	460	51.4	1.72
Tyne and Wear	228.4	**11,551**	5,313	6,238	540	50.6	1.57
NORTH WEST	**1,402.0**	**77,847**	**40,517**	**37,330**	**480**	**55.5**	**1.73**
Blackburn with Darwen UA	29.6	**2,143**	1,364	779	364	72.5	2.28
Blackpool UA	27.0	**1,549**	595	954	616	57.3	1.87
Halton UA	25.3	**1,458**	632	826	567	57.7	1.83
Warrington UA	40.0	**2,203**	1,222	981	445	55.1	1.75
Cheshire	132.6	**6,959**	4,350	2,609	375	52.5	1.64
Cumbria	90.7	**4,792**	2,646	2,146	448	52.8	1.72
Greater Manchester	541.9	**31,209**	16,327	14,882	477	57.6	1.75
Lancashire	228.8	**12,463**	6,930	5,533	444	54.5	1.74
Merseyside	286.1	**15,071**	6,451	8,620	572	52.7	1.66
YORKSHIRE AND THE HUMBER	**1,028.9**	**57,923**	**31,688**	**26,235**	**453**	**56.3**	**1.76**
East Riding of Yorkshire UA	57.4	**2,894**	1,748	1,146	396	50.4	1.68
Kingston upon Hull, City of UA	53.5	**2,970**	1,050	1,920	646	55.5	1.67
North East Lincolnshire UA	31.2	**1,748**	689	1,059	606	55.9	1.87
North Lincolnshire UA	29.3	**1,627**	771	856	526	55.4	1.86
York UA	40.5	**1,819**	1,080	739	406	44.9	1.37
North Yorkshire	104.7	**5,655**	3,630	2,025	358	54.0	1.79
South Yorkshire	262.1	**14,472**	7,103	7,369	509	55.2	1.72
West Yorkshire	450.1	**26,738**	15,617	11,121	416	59.4	1.81
EAST MIDLANDS	**861.0**	**46,916**	**26,471**	**20,445**	**436**	**54.5**	**1.70**
Derby UA	50.1	**2,935**	1,617	1,318	449	58.6	1.76
Leicester UA	67.3	**4,380**	2,627	1,753	400	65.0	1.89
Nottingham UA	66.9	**3,540**	1,529	2,011	568	52.9	1.57
Rutland UA	6.2	**314**	225	89	283	50.8	1.78
Derbyshire	143.5	**7,431**	4,161	3,270	440	51.8	1.66
Leicestershire	122.8	**6,556**	4,190	2,366	361	53.4	1.68
Lincolnshire	122.5	**6,252**	3,379	2,873	460	51.0	1.68
Northamptonshire	132.0	**7,840**	4,525	3,315	423	59.4	1.88
Nottinghamshire	149.8	**7,668**	4,218	3,450	450	51.2	1.63
WEST MIDLANDS	**1,076.5**	**63,694**	**36,994**	**26,700**	**419**	**59.2**	**1.84**
Herefordshire, County of UA	31.2	**1,668**	981	687	412	53.4	1.77
Stoke-on-Trent UA	49.8	**2,963**	1,383	1,580	533	59.5	1.83
Telford and Wrekin UA	34.0	**1,912**	940	972	508	56.2	1.78
Shropshire	50.9	**2,837**	1,775	1,062	374	55.7	1.80
Staffordshire	157.3	**8,293**	4,822	3,471	419	52.7	1.70
Warwickshire	103.3	**5,389**	3,357	2,032	377	52.2	1.60
West Midlands	544.6	**34,862**	20,235	14,627	420	64.0	1.94
Worcestershire	105.3	**5,770**	3,501	2,269	393	54.8	1.72

Note: Population estimates may not add exactly due to rounding - see section 2.15.
1 See sections 2.1 and 2.11.

Table 7.1 - *continued*

Area of usual residence of mother	Estimated number of women aged 15-44 (000s)	Live births			Live births outside marriage per 1,000 live births	General fertility rate (GFR)[1]	Total fertility rate (TFR)[1]
		Total	Within marriage	Outside marriage			
EAST	**1,087.7**	**62,711**	**38,991**	**23,720**	**378**	**57.7**	**1.77**
Luton UA	41.3	**3,085**	2,128	957	310	74.8	2.24
Peterborough UA	33.8	**2,213**	1,299	914	413	65.4	2.02
Southend-on-Sea UA	31.4	**1,911**	996	915	479	60.9	1.89
Thurrock UA	32.2	**1,982**	1,039	943	476	61.6	1.83
Bedfordshire	80.4	**4,602**	2,979	1,623	353	57.2	1.76
Cambridgeshire	120.3	**6,250**	4,127	2,123	340	51.9	1.54
Essex	256.9	**14,558**	8,909	5,649	388	56.7	1.76
Hertfordshire	217.3	**12,865**	8,676	4,189	326	59.2	1.76
Norfolk	148.2	**7,859**	4,421	3,438	437	53.0	1.67
Suffolk	125.8	**7,386**	4,417	2,969	402	58.7	1.87
LONDON	**1,807.5**	**110,437**	**72,330**	**38,107**	**345**	**61.1**	**1.71**
Inner London	786.2	**47,848**	30,327	17,521	366	60.9	1.68
Outer London	1,021.3	**62,589**	42,003	20,586	329	61.3	1.76
SOUTH EAST	**1,632.8**	**91,842**	**58,683**	**33,159**	**361**	**56.2**	**1.71**
Bracknell Forest UA	25.0	**1,425**	967	458	321	57.0	1.69
Brighton and Hove UA	60.5	**3,043**	1,562	1,481	487	50.3	1.42
Isle of Wight UA	22.9	**1,107**	566	541	489	48.2	1.61
Medway UA	53.9	**3,142**	1,662	1,480	471	58.3	1.82
Milton Keynes UA	47.7	**3,140**	1,854	1,286	410	65.8	1.99
Portsmouth UA	43.2	**2,225**	1,164	1,061	477	51.5	1.52
Reading UA	34.6	**2,013**	1,222	791	393	58.2	1.64
Slough UA	28.0	**1,983**	1,413	570	287	70.8	2.04
Southampton UA	51.9	**2,556**	1,351	1,205	471	49.3	1.44
West Berkshire UA	29.2	**1,729**	1,154	575	333	59.3	1.88
Windsor and Maidenhead UA	27.6	**1,674**	1,258	416	249	60.7	1.77
Wokingham UA	31.9	**1,653**	1,282	371	224	51.8	1.56
Buckinghamshire	95.6	**5,579**	4,088	1,491	267	58.4	1.78
East Sussex	84.9	**4,771**	2,775	1,996	418	56.2	1.87
Hampshire	245.0	**13,258**	8,668	4,590	346	54.1	1.71
Kent	262.7	**15,034**	8,700	6,334	421	57.2	1.80
Oxfordshire	133.0	**7,297**	4,996	2,301	315	54.9	1.63
Surrey	215.3	**12,304**	8,903	3,401	276	57.2	1.69
West Sussex	140.1	**7,909**	5,098	2,811	355	56.4	1.76
SOUTH WEST	**950.6**	**51,476**	**30,257**	**21,219**	**412**	**54.2**	**1.70**
Bath and North East Somerset UA	34.8	**1,633**	1,051	582	356	46.9	1.45
Bournemouth UA	33.9	**1,628**	917	711	437	48.0	1.39
Bristol, City of UA	92.9	**5,023**	2,635	2,388	475	54.1	1.57
North Somerset UA	34.2	**1,921**	1,191	730	380	56.2	1.82
Plymouth UA	50.6	**2,722**	1,300	1,422	522	53.8	1.69
Poole UA	26.1	**1,456**	872	584	401	55.8	1.78
South Gloucestershire UA	50.3	**2,870**	1,867	1,003	349	57.1	1.75
Swindon UA	39.1	**2,331**	1,393	938	402	59.6	1.83
Torbay UA	22.6	**1,267**	602	665	525	56.1	1.88
Cornwall and Isles of Scilly	90.1	**4,676**	2,562	2,114	452	51.9	1.69
Devon	125.6	**6,455**	3,864	2,591	401	51.4	1.67
Dorset	64.5	**3,352**	2,064	1,288	384	52.0	1.76
Gloucestershire	110.4	**6,088**	3,662	2,426	398	55.2	1.75
Somerset	91.0	**5,105**	2,963	2,142	420	56.1	1.85
Wiltshire	84.6	**4,949**	3,314	1,635	330	58.5	1.87

1 See sections 2.1 and 2.11.

Table 7.1 - *continued*

Area of usual residence of mother	Estimated number of women aged 15-44 (000s)	Live births			Live births outside marriage per 1,000 live births	General fertility rate (GFR)[1]	Total fertility rate (TFR)[1]
		Total	Within marriage	Outside marriage			
WALES	578.1	31,400	15,599	15,801	503	54.3	1.71
Isle of Anglesey	12.1	697	366	331	475	57.6	1.85
Gwynedd	22.0	1,156	551	605	523	52.7	1.65
Conwy	18.9	1,046	536	510	488	55.4	1.82
Denbighshire	17.0	953	442	511	536	55.9	1.85
Flintshire	29.6	1,607	902	705	439	54.2	1.72
Wrexham	26.0	1,450	722	728	502	55.8	1.72
Powys	21.8	1,153	649	504	437	53.0	1.79
Ceredigion	15.4	612	347	265	433	39.7	1.45
Pembrokeshire	20.4	1,153	582	571	495	56.5	1.86
Carmarthenshire	32.2	1,767	946	821	465	54.8	1.80
Swansea	44.5	2,432	1,231	1,201	494	54.7	1.73
Neath Port Talbot	25.9	1,368	619	749	548	52.7	1.78
Bridgend	25.6	1,473	745	728	494	57.6	1.90
The Vale of Glamorgan	23.7	1,303	677	626	480	55.0	1.81
Cardiff	75.6	3,708	2,030	1,678	453	49.0	1.47
Rhondda, Cynon, Taff	47.2	2,675	1,106	1,569	587	56.7	1.79
Merthyr Tydfil	11.1	639	222	417	653	57.4	1.92
Caerphilly	34.2	2,005	904	1,101	549	58.6	1.88
Blaenau Gwent	13.6	692	259	433	626	51.0	1.70
Torfaen	17.6	901	368	533	592	51.1	1.69
Monmouthshire	15.3	898	556	342	381	58.9	2.00
Newport	28.4	1,712	839	873	510	60.3	1.95
Normal residence outside England and Wales	:	218	160	58	266	:	:

Note: Population estimates may not add exactly due to rounding - see section 2.15.
1 See sections 2.1 and 2.11.

Table 7.2 Live births (numbers, rates, general fertility rate, and total fertility rate), and population: occurrence within/outside marriage, and area of usual residence, 2003

England and Wales, Government Office Regions (within England), and health authorities/boards

Area of usual residence of mother	Estimated number of women aged 15-44 (000s)	Live births			Live births outside marriage per 1,000 live births	General fertility rate (GFR)[1]	Total fertility rate (TFR)[1]
		Total	Within marriage	Outside marriage			
ENGLAND AND WALES	**10,944.1**	**621,469**	**364,244**	**257,225**	**414**	**56.8**	**1.73**
ENGLAND	10,366.0	589,851	348,485	241,366	409	56.9	1.73
NORTH EAST	**519.0**	**27,005**	**12,554**	**14,451**	**535**	**52.0**	**1.66**
County Durham and Tees Valley	233.6	12,520	5,656	6,864	548	53.6	1.73
Northumberland, Tyne & Wear	285.4	14,485	6,898	7,587	524	50.7	1.60
NORTH WEST	**1,402.0**	**77,847**	**40,517**	**37,330**	**480**	**55.5**	**1.73**
Cheshire & Merseyside	484.0	25,691	12,655	13,036	507	53.1	1.68
Cumbria and Lancashire	376.2	20,947	11,535	9,412	449	55.7	1.79
Greater Manchester	541.9	31,209	16,327	14,882	477	57.6	1.75
YORKSHIRE AND THE HUMBER	**1,028.9**	**57,923**	**31,688**	**26,235**	**453**	**56.3**	**1.76**
North and East Yorkshire and Northern Lincolnshire	316.7	16,713	8,968	7,745	463	52.8	1.70
South Yorkshire	262.1	14,472	7,103	7,369	509	55.2	1.72
West Yorkshire	450.1	26,738	15,617	11,121	416	59.4	1.81
EAST MIDLANDS	**861.0**	**46,916**	**26,471**	**20,445**	**436**	**54.5**	**1.70**
Leicestershire, Northamptonshire and Rutland	328.3	19,090	11,567	7,523	394	58.2	1.81
Trent	532.8	27,826	14,904	12,922	464	52.2	1.64
WEST MIDLANDS	**1,076.5**	**63,694**	**36,994**	**26,700**	**419**	**59.2**	**1.84**
Birmingham and the Black Country	479.3	31,061	18,183	12,878	415	64.8	1.97
Shropshire and Staffordshire	292.1	16,005	8,920	7,085	443	54.8	1.76
West Midlands South	305.1	16,628	9,891	6,737	405	54.5	1.69
EAST	**1,087.7**	**62,711**	**38,991**	**23,720**	**378**	**57.7**	**1.77**
Bedfordshire and Hertfordshire	339.0	20,552	13,783	6,769	329	60.6	1.83
Essex	320.5	18,451	10,944	7,507	407	57.6	1.78
Norfolk, Suffolk and Cambridgeshire	428.1	23,708	14,264	9,444	398	55.4	1.71
LONDON	**1,807.5**	**110,437**	**72,330**	**38,107**	**345**	**61.1**	**1.71**
North Central London	307.6	17,926	11,910	6,016	336	58.3	1.63
North East London	365.8	25,471	17,020	8,451	332	69.6	1.99
North West London	449.9	25,901	18,745	7,156	276	57.6	1.59
South East London	363.9	22,798	12,233	10,565	463	62.7	1.79
South West London	320.3	18,341	12,422	5,919	323	57.3	1.57
SOUTH EAST	**1,632.8**	**91,842**	**58,683**	**33,159**	**361**	**56.2**	**1.71**
Hampshire and Isle of Wight	363.0	19,146	11,749	7,397	386	52.7	1.63
Kent and Medway	316.5	18,176	10,362	7,814	430	57.4	1.81
Surrey and Sussex	500.8	28,027	18,338	9,689	346	56.0	1.69
Thames Valley	452.5	26,493	18,234	8,259	312	58.5	1.74
SOUTH WEST	**950.6**	**51,476**	**30,257**	**21,219**	**412**	**54.2**	**1.70**
Avon, Gloucestershire and Wiltshire	446.2	24,815	15,113	9,702	391	55.6	1.70
Dorset and Somerset	215.5	11,541	6,816	4,725	409	53.6	1.72
South West Peninsula	288.9	15,120	8,328	6,792	449	52.3	1.70

Note: Population estimates may not add exactly due to rounding - see section 2.15.
1 See sections 2.1 and 2.11.

Table 7.2 - *continued*

Area of usual residence of mother	Estimated number of women aged 15-44 (000s)	Live births			Live births outside marriage per 1,000 live births	General fertility rate (GFR)[1]	Total fertility rate (TFR)[1]
		Total	Within marriage	Outside marriage			
WALES	**578.1**	**31,400**	**15,599**	**15,801**	**503**	**54.3**	**1.71**
Anglesey	12.1	**697**	366	331	475	57.6	1.85
Gwynedd	22.0	**1,156**	551	605	523	52.7	1.65
Conwy	18.9	**1,046**	536	510	488	55.4	1.82
Denbighshire	17.0	**953**	442	511	536	55.9	1.85
Flintshire	29.6	**1,607**	902	705	439	54.2	1.72
Wrexham	26.0	**1,450**	722	728	502	55.8	1.72
Powys	21.8	**1,153**	649	504	437	53.0	1.79
Ceredigion	15.4	**612**	347	265	433	39.7	1.45
Pembrokeshire	20.4	**1,153**	582	571	495	56.5	1.86
Carmarthenshire	32.2	**1,767**	946	821	465	54.8	1.80
Swansea	44.5	**2,432**	1,231	1,201	494	54.7	1.73
Neath Port Talbot	25.9	**1,368**	619	749	548	52.7	1.78
Bridgend	25.6	**1,473**	745	728	494	57.6	1.90
Vale of Glamorgan	23.7	**1,303**	677	626	480	55.0	1.81
Cardiff	75.6	**3,708**	2,030	1,678	453	49.0	1.47
Rhondda Cynon Taff	47.2	**2,675**	1,106	1,569	587	56.7	1.79
Merthyr Tydfil	11.1	**639**	222	417	653	57.4	1.92
Caerphilly	34.2	**2,005**	904	1,101	549	58.6	1.88
Blaenau Gwent	13.6	**692**	259	433	626	51.0	1.70
Torfaen	17.6	**901**	368	533	592	51.1	1.69
Monmouthshire	15.3	**898**	556	342	381	58.9	2.00
Newport	28.4	**1,712**	839	873	510	60.3	1.95
Normal residence outside England and Wales	:	**218**	160	58	266	:	:

Note: Population estimates may not add exactly due to rounding - see section 2.15.
1 See sections 2.1 and 2.11.

Table 7.3 Live births (numbers and rates): age of mother and area of usual residence, 2003

England and Wales, Government Office Regions (within England), counties and unitary authorities

Area of usual residence of mother	Age of mother at birth							Age of mother at birth						
	All ages	Under 20	20-24	25-29	30-34	35-39	40 and over	All ages[1]	Under 20	20-24	25-29	30-34	35-39	40 and over
	Numbers							Rates per 1,000 women in age group[2]						
ENGLAND AND WALES	621,469	44,236	116,622	156,931	187,214	97,386	19,080	56.8	26.8	71.2	96.4	94.8	46.4	9.8
ENGLAND	589,851	41,283	109,605	148,773	178,539	93,340	18,311	56.9	26.6	70.9	96.0	95.2	46.9	9.9
NORTH EAST	27,005	2,917	6,431	6,847	7,027	3,217	566	52.0	34.4	78.4	99.2	80.1	32.7	5.8
Darlington UA	1,179	113	273	308	301	153	31	59.9	37.3	107.0	108.5	85.3	39.2	8.1
Hartlepool UA	1,065	123	316	270	240	95	21	57.9	38.8	122.4	112.8	76.0	26.5	6.0
Middlesbrough UA	1,786	235	497	476	390	162	26	59.7	42.2	96.2	121.2	82.9	31.1	4.9
Redcar and Cleveland UA	1,446	187	365	364	349	150	31	53.0	39.3	96.0	105.9	75.7	27.7	5.9
Stockton-on-Tees UA	2,115	199	476	554	613	231	42	53.8	30.7	86.6	101.8	89.8	30.7	5.6
Durham	4,929	531	1,168	1,232	1,311	592	95	49.8	33.2	75.3	98.0	76.6	30.6	5.1
Northumberland	2,934	260	570	737	852	435	80	51.4	28.0	79.4	104.6	87.1	37.3	6.6
Tyne and Wear	11,551	1,269	2,766	2,906	2,971	1,399	240	50.6	34.9	69.6	92.5	78.3	33.5	5.8
NORTH WEST	77,847	6,837	16,599	19,722	22,047	10,689	1,953	55.5	30.2	77.7	101.0	89.3	40.2	7.7
Blackburn with Darwen UA	2,143	189	599	602	494	214	45	72.5	35.3	138.4	136.7	93.8	40.2	9.2
Blackpool UA	1,549	220	385	389	345	168	42	57.3	50.6	103.7	108.9	70.1	30.8	8.4
Halton UA	1,458	128	372	381	386	168	23	57.7	29.1	102.9	104.4	88.1	37.1	4.9
Warrington UA	2,203	160	373	542	718	348	62	55.1	26.6	71.4	102.3	97.9	41.4	8.1
Cheshire	6,959	409	953	1,628	2,453	1,290	226	52.5	19.9	55.8	94.6	101.2	47.0	8.7
Cumbria	4,792	365	875	1,219	1,455	737	141	52.8	25.3	77.1	104.1	88.0	39.9	7.7
Greater Manchester	31,209	2,971	7,132	8,050	8,394	3,949	713	57.6	35.0	81.7	99.0	87.1	39.4	7.8
Lancashire	12,463	1,067	2,697	3,247	3,504	1,659	289	54.5	27.9	77.2	107.5	87.7	38.3	6.8
Merseyside	15,071	1,328	3,213	3,664	4,298	2,156	412	52.7	27.4	69.8	96.7	89.7	40.7	7.8
YORKSHIRE AND THE HUMBER	57,923	5,369	12,975	14,955	15,944	7,324	1,356	56.3	32.6	79.5	105.3	87.8	37.9	7.4
East Riding of Yorkshire UA	2,894	195	444	750	968	468	69	50.4	20.0	69.2	108.2	93.9	38.9	5.7
Kingston upon Hull, City of UA	2,970	444	804	807	617	247	51	55.5	50.7	79.9	101.8	68.9	27.0	5.9
North East Lincolnshire UA	1,748	260	497	395	409	161	26	55.9	47.9	117.5	98.8	74.5	25.9	4.4
North Lincolnshire UA	1,627	181	362	417	449	178	40	55.4	38.5	95.8	110.5	84.6	29.6	6.9
York UA	1,819	118	269	444	608	311	69	44.9	18.7	32.1	78.1	89.3	45.3	10.6
North Yorkshire	5,655	338	861	1,381	1,889	1,016	170	54.0	20.0	74.0	107.9	100.8	45.6	7.6
South Yorkshire	14,472	1,446	3,377	3,750	3,811	1,725	363	55.2	35.0	79.5	105.4	81.9	34.9	7.8
West Yorkshire	26,738	2,387	6,361	7,011	7,193	3,218	568	59.4	33.3	83.5	107.4	90.6	39.6	7.4
EAST MIDLANDS	46,916	3,564	9,238	12,137	14,004	6,769	1,204	54.5	26.6	71.5	103.0	91.0	40.2	7.6
Derby UA	2,935	287	709	780	740	359	60	58.6	37.0	80.9	103.4	84.0	39.3	7.4
Leicester UA	4,380	408	1,128	1,209	1,048	492	95	65.0	38.4	77.6	111.5	93.7	47.4	9.7
Nottingham UA	3,540	430	967	855	797	412	79	52.9	38.6	55.7	88.3	77.7	41.9	9.1
Rutland UA	314	8	50	87	106	49	14	50.8	6.1	104.4	125.9	95.8	36.8	11.2
Derbyshire	7,431	465	1,304	1,884	2,386	1,178	214	51.8	21.8	75.3	98.7	88.0	38.8	7.6
Leicestershire	6,556	292	932	1,749	2,280	1,113	190	53.4	15.1	55.0	110.4	102.4	44.9	8.1
Lincolnshire	6,252	535	1,308	1,626	1,777	840	166	51.0	26.8	79.1	104.3	82.1	34.1	6.9
Northamptonshire	7,840	574	1,445	2,060	2,427	1,151	183	59.4	28.8	81.6	113.8	99.9	42.5	7.4
Nottinghamshire	7,668	565	1,395	1,887	2,443	1,175	203	51.2	24.9	71.5	92.1	89.7	38.2	7.0
WEST MIDLANDS	63,694	5,272	13,760	16,617	17,800	8,635	1,610	59.2	30.6	83.8	108.7	92.9	42.2	8.4
Herefordshire, County of UA	1,668	122	299	418	503	254	72	53.4	23.9	82.8	105.6	90.0	38.7	11.3
Stoke-on-Trent UA	2,963	380	828	798	646	269	42	59.5	47.8	94.3	113.9	73.6	30.2	5.0
Telford and Wrekin UA	1,912	186	449	476	506	233	62	56.2	34.4	95.6	99.3	79.3	34.6	10.3
Shropshire	2,837	135	427	699	946	537	93	55.7	16.3	71.0	108.3	104.3	49.6	9.0
Staffordshire	8,293	645	1,492	2,149	2,549	1,256	202	52.7	25.6	72.6	105.2	89.5	39.3	6.6
Warwickshire	5,389	297	839	1,355	1,791	941	166	52.2	19.2	57.0	91.5	96.8	45.7	8.6
West Midlands	34,862	3,165	8,432	9,331	8,957	4,189	788	64.0	35.6	91.7	115.0	94.0	42.9	8.8
Worcestershire	5,770	342	994	1,391	1,902	956	185	54.8	21.3	72.0	97.2	97.8	44.4	9.2

1 The rates for women of all ages, under 20 and 40 and over are based on women aged 15-44, 15-19 and 40-44 respectively.

2 See sections 2.1 and 2.11.

Table 7.3 - *continued*

Area of usual residence of mother	Age of mother at birth							Age of mother at birth						
	All ages	Under 20	20-24	25-29	30-34	35-39	40 and over	All ages[1]	Under 20	20-24	25-29	30-34	35-39	40 and over
	Numbers							Rates per 1,000 women in age group[2]						
EAST	**62,711**	**3,595**	**10,475**	**16,216**	**20,255**	**10,224**	**1,946**	**57.7**	**22.1**	**69.5**	**102.9**	**102.1**	**47.2**	**9.7**
Luton UA	**3,085**	215	785	938	722	363	62	**74.8**	31.7	108.1	143.2	103.3	51.1	9.4
Peterborough UA	**2,213**	220	551	597	574	231	40	**65.4**	42.3	113.0	111.8	90.9	36.1	7.0
Southend-on-Sea UA	**1,911**	135	396	483	551	286	60	**60.9**	29.4	93.2	102.9	95.2	45.6	10.4
Thurrock UA	**1,982**	141	400	559	617	229	36	**61.6**	30.7	89.4	101.6	98.6	37.0	7.0
Bedfordshire	**4,602**	233	713	1,228	1,553	734	141	**57.2**	19.7	67.1	106.2	104.3	45.1	9.2
Cambridgeshire	**6,250**	311	943	1,514	2,149	1,137	196	**51.9**	17.6	48.7	84.7	99.5	50.0	9.3
Essex	**14,558**	765	2,158	3,807	4,937	2,407	484	**56.7**	19.7	63.6	105.8	105.6	45.8	9.9
Hertfordshire	**12,865**	555	1,555	3,013	4,627	2,631	484	**59.2**	18.2	55.3	94.8	113.2	58.9	11.7
Norfolk	**7,859**	550	1,612	2,054	2,318	1,116	209	**53.0**	23.4	74.9	100.0	88.1	38.9	7.6
Suffolk	**7,386**	470	1,362	2,023	2,207	1,090	234	**58.7**	24.1	83.2	114.4	97.5	42.6	9.8
LONDON	**110,437**	**5,272**	**17,385**	**27,193**	**34,621**	**21,162**	**4,804**	**61.1**	**24.5**	**63.3**	**77.1**	**97.5**	**63.8**	**17.2**
Inner London	**47,848**	2,498	7,981	11,217	14,448	9,439	2,265	**60.9**	30.9	62.4	63.6	89.8	70.2	21.5
Outer London	**62,589**	2,774	9,404	15,976	20,173	11,723	2,539	**61.3**	20.7	64.2	90.7	103.8	59.5	14.7
SOUTH EAST	**91,842**	**5,066**	**14,070**	**22,189**	**30,465**	**16,796**	**3,256**	**56.2**	**20.8**	**60.0**	**95.1**	**104.3**	**51.8**	**10.7**
Bracknell Forest UA	**1,425**	75	182	358	494	271	45	**57.0**	22.4	57.7	93.6	100.7	52.4	9.8
Brighton and Hove UA	**3,043**	165	394	604	1,066	677	137	**50.3**	22.1	34.2	58.2	92.5	63.4	15.4
Isle of Wight UA	**1,107**	96	253	257	302	169	30	**48.2**	24.5	88.6	88.9	76.3	35.9	6.5
Medway UA	**3,142**	258	689	872	856	385	82	**58.3**	30.3	88.1	110.6	89.2	37.4	8.4
Milton Keynes UA	**3,140**	227	573	893	937	432	78	**65.8**	33.0	89.4	116.1	102.1	47.2	9.3
Portsmouth UA	**2,225**	157	503	644	611	266	44	**51.5**	23.9	58.1	94.7	84.6	35.9	6.7
Reading UA	**2,013**	149	359	509	598	337	61	**58.2**	32.6	50.4	77.4	95.1	62.1	13.3
Slough UA	**1,983**	114	433	644	504	242	46	**70.8**	30.1	105.5	121.3	93.1	48.7	10.4
Southampton UA	**2,556**	283	574	752	598	294	55	**49.3**	37.6	42.4	86.7	74.3	39.4	8.3
West Berkshire UA	**1,729**	76	218	420	652	303	60	**59.3**	16.4	69.7	102.6	119.6	49.3	10.5
Windsor and Maidenhead UA	**1,674**	47	188	332	632	402	73	**60.7**	12.8	55.2	77.2	121.2	70.6	13.7
Wokingham UA	**1,653**	51	135	358	650	400	59	**51.8**	10.4	32.5	82.0	117.3	60.5	9.3
Buckinghamshire	**5,579**	176	692	1,274	1,999	1,203	235	**58.4**	12.2	58.3	96.9	115.4	61.3	12.2
East Sussex	**4,771**	347	803	1,170	1,406	860	185	**56.2**	24.5	76.3	114.0	98.7	48.3	10.3
Hampshire	**13,258**	727	2,010	3,279	4,404	2,379	459	**54.1**	19.4	63.7	100.6	100.6	47.0	9.4
Kent	**15,034**	1,022	2,715	3,726	4,757	2,383	431	**57.2**	24.5	75.7	104.0	101.7	45.1	8.7
Oxfordshire	**7,297**	375	942	1,680	2,524	1,461	315	**54.9**	19.4	42.2	86.1	107.1	58.2	13.6
Surrey	**12,304**	356	1,191	2,528	4,778	2,873	578	**57.2**	11.8	41.3	84.0	123.0	63.6	13.7
West Sussex	**7,909**	365	1,216	1,889	2,697	1,459	283	**56.4**	17.6	68.1	99.4	106.6	49.8	10.2
SOUTH WEST	**51,476**	**3,391**	**8,672**	**12,897**	**16,376**	**8,524**	**1,616**	**54.2**	**22.7**	**64.5**	**100.2**	**97.6**	**45.1**	**8.9**
Bath and North East Somerset UA	**1,633**	83	167	343	632	359	49	**46.9**	14.7	25.9	73.1	110.5	56.8	8.2
Bournemouth UA	**1,628**	95	302	423	482	271	55	**48.0**	20.7	44.5	75.2	82.0	46.9	10.4
Bristol, City of UA	**5,023**	409	913	1,226	1,465	855	155	**54.1**	31.1	46.6	77.7	91.4	57.4	11.5
North Somerset UA	**1,921**	121	300	454	643	346	57	**56.2**	22.9	73.9	105.2	105.8	46.6	8.1
Plymouth UA	**2,722**	285	644	770	687	282	54	**53.8**	34.4	71.5	113.9	79.9	30.5	6.2
Poole UA	**1,456**	63	283	380	443	245	42	**55.8**	15.1	84.1	105.1	94.1	47.3	8.3
South Gloucestershire UA	**2,870**	98	389	786	1,037	468	92	**57.1**	13.9	63.7	110.1	108.1	43.3	9.6
Swindon UA	**2,331**	155	406	663	744	318	45	**59.6**	28.5	81.7	105.8	101.4	40.4	6.2
Torbay UA	**1,267**	113	267	341	346	176	24	**56.1**	29.7	92.1	115.8	91.2	37.9	5.3
Cornwall and Isles of Scilly	**4,676**	357	875	1,141	1,453	692	158	**51.9**	24.0	76.4	97.9	90.0	38.3	8.8
Devon	**6,455**	395	1,019	1,513	2,144	1,140	244	**51.4**	18.9	58.2	100.7	100.5	44.8	9.6
Dorset	**3,352**	206	521	811	1,091	610	113	**52.0**	18.7	72.6	108.0	98.0	44.8	8.1
Gloucestershire	**6,088**	391	975	1,486	1,965	1,060	211	**55.2**	22.3	66.5	102.7	100.9	46.7	9.8
Somerset	**5,105**	337	878	1,322	1,611	811	146	**56.1**	22.2	81.1	113.6	99.2	42.9	8.0
Wiltshire	**4,949**	283	733	1,238	1,633	891	171	**58.5**	22.2	77.0	110.2	103.8	48.9	10.0

1 The rates for women of all ages, under 20 and 40 and over are based on women aged 15-44, 15-19 and 40-44 respectively.
2 See sections 2.1 and 2.11.

Table 7.3 - *continued*

Area of usual residence of mother	Age of mother at birth							Age of mother at birth						
	All ages	Under 20	20-24	25-29	30-34	35-39	40 and over	All ages[1]	Under 20	20-24	25-29	30-34	35-39	40 and over
	Numbers							Rates per 1,000 women in age group[2]						
WALES	**31,400**	**2,945**	**6,976**	**8,109**	**8,611**	**3,998**	**761**	**54.3**	**30.8**	**77.1**	**103.3**	**86.7**	**36.7**	**7.2**
Isle of Anglesey	**697**	58	164	180	193	83	19	**57.6**	28.1	101.0	106.8	89.2	36.0	*8.4*
Gwynedd	**1,156**	105	235	306	302	168	40	**52.7**	29.2	57.8	105.0	81.3	43.4	10.5
Conwy	**1,046**	84	194	243	321	174	30	**55.4**	27.4	81.8	99.9	96.8	46.0	7.7
Denbighshire	**953**	91	204	248	247	135	28	**55.9**	32.1	95.9	106.5	83.7	39.6	8.3
Flintshire	**1,607**	103	288	433	503	233	47	**54.2**	22.6	76.4	106.0	89.8	38.6	8.4
Wrexham	**1,450**	114	323	344	443	191	35	**55.8**	29.1	80.9	96.3	91.9	38.2	7.5
Powys	**1,153**	77	196	294	358	196	32	**53.0**	21.2	79.8	107.1	93.5	43.0	7.0
Ceredigion	**612**	48	109	162	175	97	21	**39.7**	16.2	26.5	97.4	90.0	41.1	8.8
Pembrokeshire	**1,153**	109	274	274	293	168	35	**56.5**	31.5	102.6	100.3	83.4	42.3	8.6
Carmarthenshire	**1,767**	160	378	458	491	229	51	**54.8**	28.8	84.1	108.0	90.7	36.7	8.1
Swansea	**2,432**	215	575	634	678	282	48	**54.7**	28.2	71.8	107.2	94.0	35.6	6.1
Neath Port Talbot	**1,368**	135	355	401	339	111	27	**52.7**	30.9	103.6	114.2	77.4	22.0	5.2
Bridgend	**1,473**	156	306	425	402	150	34	**57.6**	37.8	96.9	120.5	84.3	29.6	6.9
The Vale of Glamorgan	**1,303**	102	250	340	376	193	42	**55.0**	25.4	81.9	107.3	93.8	40.7	9.0
Cardiff	**3,708**	312	755	922	1,080	544	95	**49.0**	26.8	43.0	79.8	91.0	45.5	8.6
Rhondda, Cynon, Taff	**2,675**	340	683	721	653	230	48	**56.7**	43.5	90.6	111.5	77.7	26.3	5.9
Merthyr Tydfil	**639**	96	190	145	143	58	7	**57.4**	48.9	123.9	103.8	73.2	26.6	*3.3*
Caerphilly	**2,005**	207	525	565	493	188	27	**58.6**	37.0	110.1	114.4	77.4	29.0	4.4
Blaenau Gwent	**692**	106	188	176	148	64	10	**51.0**	44.7	100.2	100.6	62.0	23.7	*4.0*
Torfaen	**901**	113	241	219	198	102	28	**51.1**	37.7	101.9	90.6	65.0	29.7	8.3
Monmouthshire	**898**	45	130	206	297	192	28	**58.9**	17.7	81.6	118.7	112.1	55.0	8.6
Newport	**1,712**	169	413	413	478	210	29	**60.3**	35.0	104.4	111.6	95.1	36.7	5.6
Normal residence outside England and Wales	**218**	8	41	49	64	48	8	:	:	:	:	:	:	:

1 The rates for women of all ages, under 20 and 40 and over are based on women aged 15-44, 15-19 and 40-44 respectively.

2 See sections 2.1 and 2.11.

Table 7.4   Live births (numbers and rates): age of mother and area of usual residence, 2003

England and Wales, Government Office Regions (within England), and health authorities/boards

Area of usual residence of mother	Age of mother at birth							Age of mother at birth						
	All ages	Under 20	20-24	25-29	30-34	35-39	40 and over	All ages[1]	Under 20	20-24	25-29	30-34	35-39	40 and over
	Numbers							Rates per 1,000 women in age group[2]						
ENGLAND AND WALES	621,469	44,236	116,622	156,931	187,214	97,386	19,080	**56.8**	**26.8**	71.2	**96.4**	**94.8**	46.4	**9.8**
ENGLAND	589,851	41,283	109,605	148,773	178,539	93,340	18,311	**56.9**	26.6	70.9	96.0	95.2	46.9	9.9
NORTH EAST	27,005	2,917	6,431	6,847	7,027	3,217	566	**52.0**	**34.4**	**78.4**	**99.2**	**80.1**	**32.7**	**5.8**
County Durham and Tees Valley	12,520	1,388	3,095	3,204	3,204	1,383	246	**53.6**	35.6	88.2	104.7	80.2	30.7	5.6
Northumberland, Tyne & Wear	14,485	1,529	3,336	3,643	3,823	1,834	320	**50.7**	33.5	71.1	94.8	80.1	34.3	6.0
NORTH WEST	77,847	6,837	16,599	19,722	22,047	10,689	1,953	**55.5**	**30.2**	**77.7**	**101.0**	**89.3**	**40.2**	**7.7**
Cheshire & Merseyside	25,691	2,025	4,911	6,215	7,855	3,962	723	**53.1**	25.5	68.3	97.0	93.7	42.4	7.9
Cumbria and Lancashire	20,947	1,841	4,556	5,457	5,798	2,778	517	**55.7**	29.5	83.9	109.4	87.0	38.3	7.3
Greater Manchester	31,209	2,971	7,132	8,050	8,394	3,949	713	**57.6**	35.0	81.7	99.0	87.1	39.4	7.8
YORKSHIRE AND THE HUMBER	57,923	5,369	12,975	14,955	15,944	7,324	1,356	**56.3**	**32.6**	**79.5**	**105.3**	**87.8**	**37.9**	**7.4**
North and East Yorkshire and Northern Lincolnshire	16,713	1,536	3,237	4,194	4,940	2,381	425	**52.8**	29.7	72.8	102.0	88.8	38.0	7.0
South Yorkshire	14,472	1,446	3,377	3,750	3,811	1,725	363	**55.2**	35.0	79.5	105.4	81.9	34.9	7.8
West Yorkshire	26,738	2,387	6,361	7,011	7,193	3,218	568	**59.4**	33.3	83.5	107.4	90.6	39.6	7.4
EAST MIDLANDS	46,916	3,564	9,238	12,137	14,004	6,769	1,204	**54.5**	**26.6**	**71.5**	**103.0**	**91.0**	**40.2**	**7.6**
Leicestershire, Northamptonshire and Rutland	19,090	1,282	3,555	5,105	5,861	2,805	482	**58.2**	25.0	71.6	112.3	99.6	44.1	8.1
Trent	27,826	2,282	5,683	7,032	8,143	3,964	722	**52.2**	27.5	71.5	97.1	85.7	37.9	7.4
WEST MIDLANDS	63,694	5,272	13,760	16,617	17,800	8,635	1,610	**59.2**	**30.6**	**83.8**	**108.7**	**92.9**	**42.2**	**8.4**
Birmingham and the Black Country	31,061	2,810	7,540	8,329	7,939	3,744	699	**64.8**	36.1	94.8	116.9	94.0	43.2	8.8
Shropshire and Staffordshire	16,005	1,346	3,196	4,122	4,647	2,295	399	**54.8**	28.8	79.8	106.5	88.2	39.3	7.2
West Midlands South	16,628	1,116	3,024	4,166	5,214	2,596	512	**54.5**	23.5	67.9	97.0	95.9	43.4	9.1
EAST	62,711	3,595	10,475	16,216	20,255	10,224	1,946	**57.7**	**22.1**	**69.5**	**102.9**	**102.1**	**47.2**	**9.7**
Bedfordshire and Hertfordshire	20,552	1,003	3,053	5,179	6,902	3,728	687	**60.6**	20.4	66.4	103.8	110.0	54.8	10.8
Essex	18,451	1,041	2,954	4,849	6,105	2,922	580	**57.6**	21.7	69.2	105.0	103.8	45.0	9.7
Norfolk, Suffolk and Cambridgeshire	23,708	1,551	4,468	6,188	7,248	3,574	679	**55.4**	23.5	71.9	100.7	94.3	42.8	8.7
LONDON	110,437	5,272	17,385	27,193	34,621	21,162	4,804	**61.1**	**24.5**	**63.3**	**77.1**	**97.5**	**63.8**	**17.2**
North Central London	17,926	880	2,663	4,174	5,602	3,691	916	**58.3**	24.5	54.6	68.0	93.7	66.6	19.8
North East London	25,471	1,416	5,425	7,269	6,854	3,705	802	**69.6**	28.5	94.3	105.6	98.6	57.1	14.5
North West London	25,901	962	3,655	6,426	8,445	5,172	1,241	**57.6**	18.5	50.9	71.1	94.9	64.9	18.5
South East London	22,798	1,354	3,492	5,447	7,074	4,419	1,012	**62.7**	31.6	67.8	81.0	98.1	62.7	17.0
South West London	18,341	660	2,150	3,877	6,646	4,175	833	**57.3**	18.9	48.0	60.0	102.7	68.6	16.5
SOUTH EAST	91,842	5,066	14,070	22,189	30,465	16,796	3,256	**56.2**	**20.8**	**60.0**	**95.1**	**104.3**	**51.8**	**10.7**
Hampshire and Isle of Wight	19,146	1,263	3,340	4,932	5,915	3,108	588	**52.7**	22.7	59.0	96.8	93.9	44.3	8.8
Kent and Medway	18,176	1,280	3,404	4,598	5,613	2,768	513	**57.4**	25.5	77.9	105.2	99.5	43.9	8.6
Surrey and Sussex	28,027	1,233	3,604	6,191	9,947	5,869	1,183	**56.0**	17.0	52.5	88.8	110.6	57.0	12.2
Thames Valley	26,493	1,290	3,722	6,468	8,990	5,051	972	**58.5**	19.7	56.7	94.0	108.5	57.5	11.9
SOUTH WEST	51,476	3,391	8,672	12,897	16,376	8,524	1,616	**54.2**	**22.7**	**64.5**	**100.2**	**97.6**	**45.1**	**8.9**
Avon, Gloucestershire and Wiltshire	24,815	1,540	3,883	6,196	8,119	4,297	780	**55.6**	23.0	59.4	97.0	101.5	48.7	9.5
Dorset and Somerset	11,541	701	1,984	2,936	3,627	1,937	356	**53.6**	20.1	70.5	103.4	95.6	44.6	8.4
South West Peninsula	15,120	1,150	2,805	3,765	4,630	2,290	480	**52.3**	24.0	68.6	103.5	92.8	39.9	8.5

1 The rates for women of all ages, under 20 and 40 and over are based on women aged 15-44, 15-19 and 40-44 respectively.
2 See sections 2.1 and 2.11.

Table 7.4 - *continued*

Area of usual residence of mother	Age of mother at birth							Age of mother at birth						
	All ages	Under 20	20-24	25-29	30-34	35-39	40 and over	All ages[1]	Under 20	20-24	25-29	30-34	35-39	40 and over
	Numbers							Rates per 1,000 women in age group[2]						
WALES	**31,400**	**2,945**	**6,976**	**8,109**	**8,611**	**3,998**	**761**	**54.3**	**30.8**	**77.1**	**103.3**	**86.7**	**36.7**	**7.2**
Anglesey	**697**	58	164	180	193	83	19	**57.6**	28.1	101.0	106.8	89.2	36.0	*8.4*
Gwynedd	**1,156**	105	235	306	302	168	40	**52.7**	29.2	57.8	105.0	81.3	43.4	10.5
Conwy	**1,046**	84	194	243	321	174	30	**55.4**	27.4	81.8	99.9	96.8	46.0	7.7
Denbighshire	**953**	91	204	248	247	135	28	**55.9**	32.1	95.9	106.5	83.7	39.6	8.3
Flintshire	**1,607**	103	288	433	503	233	47	**54.2**	22.6	76.4	106.0	89.8	38.6	8.4
Wrexham	**1,450**	114	323	344	443	191	35	**55.8**	29.1	80.9	96.3	91.9	38.2	7.5
Powys	**1,153**	77	196	294	358	196	32	**53.0**	21.2	79.8	107.1	93.5	43.0	7.0
Ceredigion	**612**	48	109	162	175	97	21	**39.7**	16.2	26.5	97.4	90.0	41.1	8.8
Pembrokeshire	**1,153**	109	274	274	293	168	35	**56.5**	31.5	102.6	100.3	83.4	42.3	8.6
Carmarthenshire	**1,767**	160	378	458	491	229	51	**54.8**	28.8	84.1	108.0	90.7	36.7	8.1
Swansea	**2,432**	215	575	634	678	282	48	**54.7**	28.2	71.8	107.2	94.0	35.6	6.1
Neath Port Talbot	**1,368**	135	355	401	339	111	27	**52.7**	30.9	103.6	114.2	77.4	22.0	5.2
Bridgend	**1,473**	156	306	425	402	150	34	**57.6**	37.8	96.9	120.5	84.3	29.6	6.9
Vale of Glamorgan	**1,303**	102	250	340	376	193	42	**55.0**	25.4	81.9	107.3	93.8	40.7	9.0
Cardiff	**3,708**	312	755	922	1,080	544	95	**49.0**	26.8	43.0	79.8	91.0	45.5	8.6
Rhondda Cynon Taff	**2,675**	340	683	721	653	230	48	**56.7**	43.5	90.6	111.5	77.7	26.3	5.9
Merthyr Tydfil	**639**	96	190	145	143	58	7	**57.4**	48.9	123.9	103.8	73.2	26.6	*3.3*
Caerphilly	**2,005**	207	525	565	493	188	27	**58.6**	37.0	110.1	114.4	77.4	29.0	4.4
Blaenau Gwent	**692**	106	188	176	148	64	10	**51.0**	44.7	100.2	100.6	62.0	23.7	*4.0*
Torfaen	**901**	113	241	219	198	102	28	**51.1**	37.7	101.9	90.6	65.0	29.7	8.3
Monmouthshire	**898**	45	130	206	297	192	28	**58.9**	17.7	81.6	118.7	112.1	55.0	8.6
Newport	**1,712**	169	413	413	478	210	29	**60.3**	35.0	104.4	111.6	95.1	36.7	5.6
Normal residence outside England and Wales	**218**	8	41	49	64	48	8	:	:	:	:	:	:	:

1 The rates for women of all ages, under 20 and 40 and over are based on women aged 15-44, 15-19 and 40-44 respectively.
2 See sections 2.1 and 2.11.

Table 7.5 Live births (numbers and percentages): birthweight and area of usual residence, 2003

England and Wales, Government Office Regions (within England)

Area of usual residence of mother	Birthweight (grams)													
	All weights[1]	Under 1,500	1,500-1,999	2,000-2,499	2,500-2,999	3,000-3,499	3,500 and over	All weights[1]	Under 1,500	1,500-1,999	2,000-2,499	2,500-2,999	3,000-3,499	3,500 and over
	Numbers							**Percentages**						
ENGLAND AND WALES	**621,469**	**7,933**	**9,647**	**29,996**	**107,005**	**221,423**	**244,546**	**100**	**1.3**	**1.6**	**4.8**	**17.2**	**35.6**	**39.3**
ENGLAND	**589,851**	**7,527**	**9,166**	**28,558**	**101,700**	**210,148**	**231,853**	**100**	**1.3**	**1.6**	**4.8**	**17.2**	**35.6**	**39.3**
North East	**27,005**	350	462	1,324	4,519	9,386	10,946	**100**	1.3	1.7	4.9	16.7	34.8	40.5
North West	**77,847**	1,014	1,299	4,063	13,533	27,496	30,342	**100**	1.3	1.7	5.2	17.4	35.3	39.0
Yorkshire and the Humber	**57,923**	776	982	3,026	10,443	20,431	22,214	**100**	1.3	1.7	5.2	18.0	35.3	38.4
East Midlands	**46,916**	599	745	2,332	8,136	16,476	18,615	**100**	1.3	1.6	5.0	17.3	35.1	39.7
West Midlands	**63,694**	985	1,067	3,392	11,907	23,037	23,193	**100**	1.5	1.7	5.3	18.7	36.2	36.4
East	**62,711**	687	897	2,693	10,021	22,024	26,345	**100**	1.1	1.4	4.3	16.0	35.1	42.0
London	**110,437**	1,500	1,743	5,681	21,021	40,914	39,110	**100**	1.4	1.6	5.1	19.0	37.0	35.4
South East	**91,842**	1,047	1,257	3,860	14,198	32,357	39,049	**100**	1.1	1.4	4.2	15.5	35.2	42.5
South West	**51,476**	569	714	2,187	7,922	18,027	22,039	**100**	1.1	1.4	4.2	15.4	35.0	42.8
WALES	**31,400**	391	465	1,424	5,267	11,213	12,621	**100**	1.2	1.5	4.5	16.8	35.7	40.2
Normal residence outside England and Wales	218	15	16	14	38	62	72	**100**	6.9	7.3	6.4	17.4	28.4	33.0

1 'All weights' includes births where the birthweight was not stated.

Table 7.6 Stillbirths (numbers and rates): occurrence within/outside marriage, and area of usual residence, 2003

England and Wales, Government Office Regions (within England)

Area of usual residence of mother	Stillbirths			Stillbirths per 1,000 live and stillbirths[1]
	Total	Within marriage	Outside marriage	
ENGLAND AND WALES	**3,585**	**1,961**	**1,624**	**5.7**
ENGLAND	**3,405**	**1,870**	**1,535**	**5.7**
North East	**150**	69	81	5.5
North West	**468**	236	232	6.0
Yorkshire and the Humber	**347**	171	176	6.0
East Midlands	**289**	157	132	6.1
West Midlands	**387**	223	164	6.0
East	**309**	173	136	4.9
London	**752**	450	302	6.8
South East	**443**	254	189	4.8
South West	**260**	137	123	5.0
WALES	**160**	79	81	5.1
Normal residence outside England and Wales	**20**	12	8	84.0

1 See section 2.11.

Table 7.7 Stillbirths (numbers and percentages): birthweight and area of usual residence, 2003

England and Wales, Government Office Regions (within England)

Area of usual residence of mother	Birthweight (grams)												
	All weights[1]	Under 1,500	1,500-1,999	2,000-2,499	2,500-2,999	3,000-3,499	3,500 and over	All weights[1]	Under 1,500	1,500-1,999	2,000-2,499	2,500-2,999	3,000-3,499	3,500 and over
	Numbers							Percentages						
ENGLAND AND WALES	**3,585**	**1,647**	**364**	**400**	**406**	**404**	**279**	**100**	**45.9**	**10.2**	**11.2**	**11.3**	**11.3**	**7.8**
ENGLAND	**3,405**	**1,563**	**341**	**377**	**387**	**386**	**271**	**100**	**45.9**	**10.0**	**11.1**	**11.4**	**11.3**	**8.0**
North East	**150**	65	19	23	17	15	10	**100**	43.3	12.7	15.3	*11.3*	*10.0*	*6.7*
North West	**468**	210	42	64	57	48	33	**100**	44.9	9.0	13.7	12.2	10.3	7.1
Yorkshire and the Humber	**347**	155	41	34	46	32	29	**100**	44.7	11.8	9.8	13.3	9.2	8.4
East Midlands	**289**	133	40	25	26	33	25	**100**	46.0	13.8	8.7	9.0	11.4	8.7
West Midlands	**387**	191	29	49	37	45	31	**100**	49.4	7.5	12.7	9.6	11.6	8.0
East	**309**	144	33	30	34	32	28	**100**	46.6	10.7	9.7	11.0	10.4	9.1
London	**752**	356	78	87	76	80	58	**100**	47.3	10.4	11.6	10.1	10.4	7.7
South East	**443**	197	42	42	57	58	35	**100**	44.5	9.5	9.5	12.9	13.1	7.9
South West	**260**	112	17	23	37	43	22	**100**	43.1	6.5	8.8	14.2	16.5	8.5
WALES	**160**	71	20	20	19	18	8	**100**	44.4	12.5	12.5	*11.9*	*11.3*	*5.0*
Normal residence outside England and Wales	**20**	13	3	3	-	-	-	**100**	*65.0*	*15.0*	*15.0*	-	-	-

1 'All weights' includes births where the birthweight was not stated.

Table 8.1 Maternities: age of mother, occurrence within/outside marriage, number of previous live-born children[1] and place of confinement, 2003   **England and Wales**

Place of confinement[2]		Age of mother at birth							
		All ages	Under 20	20-24	25-29	30-34	35-39	40-44	45 and over
Total	**Total**	**615,787**	**44,245**	**116,147**	**155,802**	**184,935**	**95,841**	**17,998**	**819**
	NHS Hospitals	**598,536**	**43,829**	**114,150**	**152,063**	**178,698**	**91,734**	**17,277**	**785**
	Non-NHS Hospitals	**2,877**	**27**	**259**	**503**	**1,090**	**801**	**184**	**13**
	At Home	**13,590**	**305**	**1,545**	**3,032**	**4,959**	**3,216**	**514**	**19**
	Elsewhere	**784**	**84**	**193**	**204**	**188**	**90**	**23**	**2**
All born within marriage	**Total**	**360,114**	**4,346**	**40,684**	**97,931**	**136,181**	**68,453**	**11,946**	**573**
	NHS Hospitals	348,621	4,290	39,844	95,551	131,514	65,442	11,433	547
	Non-NHS Hospitals	2,617	22	228	468	1,013	710	165	11
	At Home	8,560	29	564	1,821	3,534	2,257	341	14
	Elsewhere	316	5	48	91	120	44	7	1
Previous live-born 0	**Total**	**148,743**	**3,477**	**22,061**	**47,949**	**53,151**	**19,095**	**2,823**	**187**
	NHS Hospitals	146,274	3,442	21,801	47,320	52,108	18,673	2,747	183
	Non-NHS Hospitals	1,107	18	134	246	452	206	47	4
	At Home	1,310	15	116	371	568	212	28	-
	Elsewhere	52	2	10	12	23	4	1	-
1	**Total**	**131,625**	**793**	**13,858**	**32,317**	**53,760**	**26,734**	**4,033**	**130**
	NHS Hospitals	126,736	777	13,463	31,270	51,669	25,573	3,860	124
	Non-NHS Hospitals	980	4	79	163	384	282	65	3
	At Home	3,774	10	295	844	1,655	860	108	2
	Elsewhere	135	2	21	40	52	19	-	1
2	**Total**	**51,519**	**66**	**3,820**	**12,028**	**19,093**	**13,932**	**2,501**	**79**
	NHS Hospitals	48,842	62	3,675	11,561	18,063	13,033	2,375	73
	Non-NHS Hospitals	375	-	10	50	129	154	28	4
	At Home	2,222	4	124	392	873	733	94	2
	Elsewhere	80	-	11	25	28	12	4	-
3	**Total**	**17,496**	**9**	**774**	**4,045**	**6,368**	**5,009**	**1,235**	**56**
	NHS Hospitals	16,517	8	740	3,877	6,019	4,662	1,158	53
	Non-NHS Hospitals	99	-	3	7	32	43	14	-
	At Home	848	-	25	152	307	300	61	3
	Elsewhere	32	1	6	9	10	4	2	-
4	**Total**	**6,128**	**-**	**133**	**1,149**	**2,347**	**1,860**	**592**	**47**
	NHS Hospitals	5,818	-	127	1,097	2,240	1,752	559	43
	Non-NHS Hospitals	31	-	2	2	11	11	5	-
	At Home	269	-	4	47	93	93	28	4
	Elsewhere	10	-	-	3	3	4	-	-
4 and over	**Total**	**10,731**	**1**	**171**	**1,592**	**3,809**	**3,683**	**1,354**	**121**
	NHS Hospitals	10,252	1	165	1,523	3,655	3,501	1,293	114
	Non-NHS Hospitals	56	-	2	2	16	25	11	-
	At Home	406	-	4	62	131	152	50	7
	Elsewhere	17	-	-	5	7	5	-	-
5-9	**Total**	**4,467**	**1**	**36**	**439**	**1,449**	**1,769**	**709**	**64**
	NHS Hospitals	4,301	1	36	422	1,403	1,697	681	61
	Non-NHS Hospitals	25	-	-	-	5	14	6	-
	At Home	134	-	-	15	37	57	22	3
	Elsewhere	7	-	-	2	4	1	-	-
10-14	**Total**	**135**	**-**	**2**	**4**	**13**	**54**	**52**	**10**
	NHS Hospitals	132	-	2	4	12	52	52	10
	Non-NHS Hospitals	-	-	-	-	-	-	-	-
	At Home	3	-	-	-	1	2	-	-
	Elsewhere	-	-	-	-	-	-	-	-
15 and over	**Total**	**1**	**-**	**-**	**-**	**-**	**-**	**1**	**-**
	NHS Hospitals	1	-	-	-	-	-	1	-
	Non-NHS Hospitals	-	-	-	-	-	-	-	-
	At Home	-	-	-	-	-	-	-	-
	Elsewhere	-	-	-	-	-	-	-	-
All born outside marriage	**Total**	**255,673**	**39,899**	**75,463**	**57,871**	**48,754**	**27,388**	**6,052**	**246**
	NHS Hospitals	249,915	39,539	74,306	56,512	47,184	26,292	5,844	238
	Non-NHS Hospitals	260	5	31	35	77	91	19	2
	At Home	5,030	276	981	1,211	1,425	959	173	5
	Elsewhere	468	79	145	113	68	46	16	1

1 See section 2.9.
2 See section 3.7.

Table 8.2 Maternities: place of confinement and whether area of occurrence is the same as area of usual residence, or other than area of usual residence, 2003

England and Wales, Government Office Regions (within England), and health authorities/boards

Area of usual residence	Place of confinement[1]					Health authority/board of occurrence		
	Total	NHS hospitals	Non-NHS hospitals	At home	Elsewhere	Same as usual residence	Other than usual residence	At home and elsewhere
ENGLAND AND WALES	**615,787**	**598,536**	**2,877**	**13,590**	**784**	**351,230**	**264,557**	**-**
ENGLAND	**584,450**	568,141	2,837	12,746	726	332,621	251,829	-
NORTH EAST	**26,764**	**26,458**	**1**	**280**	**25**	**26,376**	**83**	**305**
County Durham and Tees Valley	**12,399**	12,260	-	125	14	11,618	642	139
Northumberland, Tyne & Wear	**14,365**	14,198	1	155	11	14,067	132	166
NORTH WEST	**77,184**	**76,018**	**4**	**1,083**	**79**	**75,485**	**537**	**1,162**
Cheshire & Merseyside	**25,400**	25,095	1	265	39	23,512	1,584	304
Cumbria and Lancashire	**20,795**	20,414	1	357	23	19,409	1,006	380
Greater Manchester	**30,989**	30,509	2	461	17	30,106	405	478
YORKSHIRE AND THE HUMBER	**57,496**	**56,440**	**8**	**952**	**96**	**55,770**	**678**	**1,048**
North and East Yorkshire and Northern Lincolnshire	**16,567**	16,232	3	297	35	15,282	953	332
South Yorkshire	**14,379**	14,089	3	254	33	13,699	393	287
West Yorkshire	**26,550**	26,119	2	401	28	25,645	476	429
EAST MIDLANDS	**46,510**	**45,447**	**10**	**997**	**56**	**40,575**	**4,882**	**1,053**
Leicestershire, Northamptonshire and Rutland	**18,923**	18,447	8	439	29	15,901	2,554	468
Trent	**27,587**	27,000	2	558	27	24,123	2,879	585
WEST MIDLANDS	**63,147**	**62,084**	**8**	**978**	**77**	**61,127**	**965**	**1,055**
Birmingham and the Black Country	**30,870**	30,443	3	392	32	30,217	229	424
Shropshire and Staffordshire	**15,845**	15,556	-	259	30	13,392	2,164	289
West Midlands South	**16,432**	16,085	5	327	15	15,071	1,019	342
EAST	**62,075**	**59,496**	**632**	**1,864**	**83**	**56,703**	**3,425**	**1,947**
Bedfordshire and Hertfordshire	**20,358**	19,730	72	534	22	16,486	3,316	556
Essex	**18,253**	17,592	48	597	16	15,718	1,922	613
Norfolk, Suffolk and Cambridgeshire	**23,464**	22,174	512	733	45	22,504	182	778
LONDON	**109,438**	**105,337**	**1,997**	**2,008**	**96**	**105,534**	**1,800**	**2,104**
North Central London	**17,714**	16,785	617	299	13	15,079	2,323	312
North East London	**25,341**	24,862	90	367	22	23,398	1,554	389
North West London	**25,659**	24,384	1,011	249	15	23,638	1,757	264
South East London	**22,615**	21,788	88	710	29	20,530	1,346	739
South West London	**18,109**	17,518	191	383	17	13,647	4,062	400
SOUTH EAST	**90,870**	**87,922**	**158**	**2,667**	**123**	**85,030**	**3,050**	**2,790**
Hampshire and Isle of Wight	**18,960**	18,366	5	557	32	15,293	3,078	589
Kent and Medway	**18,004**	17,398	28	557	21	16,768	658	578
Surrey and Sussex	**27,686**	26,687	64	902	33	25,068	1,683	935
Thames Valley	**26,220**	25,471	61	651	37	23,573	1,959	688
SOUTH WEST	**50,966**	**48,939**	**19**	**1,917**	**91**	**48,498**	**460**	**2,008**
Avon, Gloucestershire and Wiltshire	**24,532**	23,818	9	669	36	23,373	454	705
Dorset and Somerset	**11,435**	10,981	6	434	14	9,931	1,056	448
South West Peninsula	**14,999**	14,140	4	814	41	13,981	163	855
WALES	**31,115**	**30,219**	**1**	**844**	**51**	**29,042**	**1,178**	**895**
Anglesey	**686**	673	-	11	2	-	673	13
Gwynedd	**1,143**	1,096	-	43	4	963	133	47
Conwy	**1,035**	1,018	-	17	-	-	1,018	17
Denbighshire	**946**	923	-	22	1	825	98	23
Flintshire	**1,584**	1,568	1	13	2	-	1,569	15
Wrexham	**1,440**	1,414	-	26	-	1,358	56	26
Powys	**1,147**	1,068	-	77	2	176	892	79
Ceredigion	**599**	571	-	27	1	360	211	28

1 See section 3.7.

Table 8.2 - *continued*

Area of usual residence	Place of confinement[1]					Health authority/board of occurrence		
	Total	NHS hospitals	Non-NHS hospitals	At home	Elsewhere	Same as usual residence	Other than usual residence	At home and elsewhere
WALES - *continued*								
Pembrokeshire	**1,146**	1,097	-	43	6	1,057	40	49
Carmarthenshire	**1,749**	1,630	-	119	-	1,131	499	119
Swansea	**2,411**	2,366	-	44	1	2,334	32	45
Neath Port Talbot	**1,354**	1,321	-	31	2	884	437	33
Bridgend	**1,465**	1,414	-	50	1	1,352	62	51
Vale of Glamorgan	**1,290**	1,252	-	37	1	1,082	170	38
Cardiff	**3,671**	3,592	-	75	4	2,202	1,390	79
Rhondda Cynon Taff	**2,659**	2,598	-	55	6	2,052	546	61
Merthyr Tydfil	**625**	620	-	4	1	580	40	5
Caerphilly	**1,993**	1,940	-	47	6	281	1,659	53
Blaenau Gwent	**684**	655	-	28	1	-	655	29
Torfaen	**897**	877	-	18	2	-	877	20
Monmouthshire	**887**	854	-	30	3	391	463	33
Newport	**1,704**	1,672	-	27	5	1,574	98	32
Normal residence outside England and Wales	**222**	176	39	-	7	7	215	-

1   See section 3.7.

Table 8.3   Maternities in hospitals: live births and stillbirths, and area of occurrence[1], 2003

England and Wales, Government Office Regions (within England), and health authorities/boards

Area of occurrence	Maternities	Live births	Stillbirths	Stillbirths per 1,000 live births and stillbirths
ENGLAND AND WALES	**601,413**	**607,121**	**3,526**	**5.8**
ENGLAND	**572,022**	577,445	3,377	5.8
NORTH EAST	**26,631**	**26,878**	**151**	**5.6**
County Durham and Tees Valley	**11,894**	12,007	66	5.5
Northumberland, Tyne & Wear	**14,737**	14,871	85	5.7
NORTH WEST	**77,385**	**78,071**	**475**	**6.1**
Cheshire & Merseyside	**24,803**	25,086	130	5.2
Cumbria and Lancashire	**20,541**	20,691	132	6.3
Greater Manchester	**32,041**	32,294	213	6.6
YORKSHIRE AND THE HUMBER	**56,821**	**57,238**	**352**	**6.1**
North and East Yorkshire and Northern Lincolnshire	**16,008**	16,142	85	5.2
South Yorkshire	**14,153**	14,247	99	6.9
West Yorkshire	**26,660**	26,849	168	6.2
EAST MIDLANDS	**41,188**	**41,570**	**246**	**5.9**
Leicestershire, Northamptonshire and Rutland	**16,073**	16,221	94	5.8
Trent	**25,115**	25,349	152	6.0
WEST MIDLANDS	**63,528**	**64,082**	**397**	**6.2**
Birmingham and the Black Country	**32,476**	32,702	239	7.3
Shropshire and Staffordshire	**15,039**	15,192	85	5.6
West Midlands South	**16,013**	16,188	73	4.5
EAST	**58,805**	**59,430**	**300**	**5.0**
Bedfordshire and Hertfordshire	**17,349**	17,520	103	5.8
Essex	**16,452**	16,641	75	4.5
Norfolk, Suffolk and Cambridgeshire	**25,004**	25,269	122	4.8
LONDON	**110,517**	**111,582**	**768**	**6.8**
North Central London	**18,024**	18,213	107	5.8
North East London	**25,197**	25,348	199	7.8
North West London	**29,592**	29,931	184	6.1
South East London	**22,330**	22,525	193	8.5
South West London	**15,374**	15,565	85	5.4
SOUTH EAST	**87,496**	**88,438**	**426**	**4.8**
Hampshire and Isle of Wight	**15,859**	16,028	83	5.2
Kent and Medway	**17,268**	17,437	92	5.3
Surrey and Sussex	**29,232**	29,582	134	4.5
Thames Valley	**25,137**	25,391	117	4.6
SOUTH WEST	**49,651**	**50,156**	**262**	**5.2**
Avon, Gloucestershire and Wiltshire	**25,257**	25,545	117	4.6
Dorset and Somerset	**10,292**	10,384	60	5.7
South West Peninsula	**14,102**	14,227	85	5.9
WALES	**29,391**	**29,676**	**149**	**5.0**
Anglesey	-	-	-	-
Gwynedd	**1,818**	1,840	8	*4.3*
Conwy	-	-	-	-
Denbighshire	**2,227**	2,252	9	*4.0*
Flintshire	-	-	-	-
Wrexham	**2,300**	2,328	5	*2.1*
Powys	**182**	182	-	-
Ceredigion	**481**	487	2	*4.1*
Pembrokeshire	**1,130**	1,139	8	*7.0*

1 See section 3.7.

Table 8.3 - *continued*

Area of occurrence	Maternities	Live births	Stillbirths	Stillbirths per 1,000 live births and stillbirths
WALES - *continued*				
Carmarthenshire	1,318	1,326	1	0.8
Swansea	3,230	3,293	10	3.0
Neath Port Talbot	951	946	5	5.3
Bridgend	1,544	1,556	10	6.4
Vale of Glamorgan	2,422	2,447	8	3.3
Cardiff	2,858	2,878	39	13.4
Rhondda Cynon Taff	2,218	2,239	12	5.3
Merthyr Tydfil	1,229	1,244	6	4.8
Caerphilly	299	299	-	-
Blaenau Gwent	-	-	-	-
Torfaen	-	-	-	-
Monmouthshire	1,613	1,628	3	1.8
Newport	3,571	3,592	23	6.4

Table 9.1 Live births (numbers and percentages): country of birth of mother[4], 1993, 1998-2003

Country of birth of mother	1993	1998	1999	2000	2001	2002	2003
	Numbers						
Total	**673,467**	**635,901**	**621,872**	**604,441**	**594,634**	**596,122**	**621,469**
United Kingdom[1]	592,056	549,432	532,852	510,835	496,713	490,711	506,076
Total outside United Kingdom	**81,367**	**86,456**	**89,000**	**93,588**	**97,895**	**105,381**	**115,360**
Irish Republic	5,149	4,673	4,470	4,050	3,843	3,708	3,734
Australia, Canada and New Zealand	3,161	3,393	3,531	3,635	3,695	3,885	4,132
New Commonwealth	**47,793**	**46,023**	**46,201**	**47,249**	**50,047**	**54,037**	**55,777**
India	7,316	6,513	6,497	6,650	6,598	7,222	7,995
Pakistan	13,023	13,069	13,472	13,561	14,588	15,357	15,107
Bangladesh	5,854	7,424	7,375	7,482	8,164	8,486	8,898
East Africa	5,698	4,498	4,159	3,959	3,745	3,724	3,985
Southern Africa	:	1,437	1,608	1,907	2,236	2,654	3,149
Rest of Africa	5,511	6,135	6,138	6,537	6,995	8,073	7,456
Caribbean	3,124	2,564	2,536	2,681	3,085	3,598	3,984
Far East[2]	3,710	1,682	1,539	1,538	1,365	1,351	1,433
Mediterranean[3]	1,994	1,369	1,268	1,148	1,099	1,065	945
Rest of New Commonwealth	1,563	1,332	1,609	1,786	2,172	2,507	2,825
Other European Union	7,850	10,414	10,845	11,105	10,944	11,449	12,152
Rest of Europe	2,863	4,522	5,635	7,362	7,649	8,142	9,475
United States of America	2,932	2,857	2,780	2,895	2,878	2,828	3,037
Rest of the World	11,619	14,574	15,538	17,292	18,839	21,332	27,053
Not stated	44	13	20	18	26	30	33
	Percentage of all live births						
Total	**100.0**	**100.0**	**100.0**	**100.0**	**100.0**	**100.0**	**100.0**
United Kingdom[1]	87.9	86.4	85.7	84.5	83.5	82.3	81.4
Total outside United Kingdom	**12.1**	**13.6**	**14.3**	**15.5**	**16.5**	**17.7**	**18.6**
Irish Republic	0.8	0.7	0.7	0.7	0.6	0.6	0.6
Australia, Canada and New Zealand	0.5	0.5	0.6	0.6	0.6	0.7	0.7
New Commonwealth	**7.1**	**7.2**	**7.4**	**7.8**	**8.4**	**9.1**	**9.0**
India	1.1	1.0	1.0	1.1	1.1	1.2	1.3
Pakistan	1.9	2.1	2.2	2.2	2.5	2.6	2.4
Bangladesh	0.9	1.2	1.2	1.2	1.4	1.4	1.4
East Africa	0.8	0.7	0.7	0.7	0.6	0.6	0.6
Southern Africa	:	0.2	0.3	0.3	0.4	0.4	0.5
Rest of Africa	0.8	1.0	1.0	1.1	1.2	1.4	1.2
Caribbean	0.5	0.4	0.4	0.4	0.5	0.6	0.6
Far East[2]	0.6	0.3	0.2	0.3	0.2	0.2	0.2
Mediterranean[3]	0.3	0.2	0.2	0.2	0.2	0.2	0.2
Rest of New Commonwealth	0.2	0.2	0.3	0.3	0.4	0.4	0.5
Other European Union	1.2	1.6	1.7	1.8	1.8	1.9	2.0
Rest of Europe	0.4	0.7	0.9	1.2	1.3	1.4	1.5
United States of America	0.4	0.4	0.4	0.5	0.5	0.5	0.5
Rest of the World	1.7	2.3	2.5	2.9	3.2	3.6	4.4
Not stated	0.0	0.0	0.0	0.0	0.0	0.0	0.0

1 Including Isle of Man and Channel Islands.
2 Brunei, Malaysia and Singapore, and Hong Kong before 1997.
3 Cyprus, Gibraltar and Malta.
4 See section 2.7.

Table 9.2 Live births (numbers and percentages): birthplace of mother if outside United Kingdom[3], and area of usual residence, 2003

England and Wales, Government Office Regions (within England): counties[2], unitary authorities, county districts and London boroughs, with more than 15 per cent non-UK born mothers

Area of usual residence of mother	All live births	Birthplace of mother outside United Kingdom					
		New Commonwealth		Rest of the World		All outside United Kingdom	
		Number	Percentage	Number	Percentage	Number	Percentage
ENGLAND AND WALES[1]	**621,251**	**55,756**	**9**	**59,548**	**10**	**115,304**	**19**
ENGLAND	589,851	54,997	9	58,262	10	113,259	19
NORTH EAST	**27,005**	**791**	**3**	**1,066**	**4**	**1,857**	**7**
Tyne and Wear (Met County)	**11,551**	457	4	577	5	1,034	9
Newcastle upon Tyne	**2,895**	220	8	261	9	481	17
NORTH WEST	**77,847**	**5,167**	**7**	**3,745**	**5**	**8,912**	**11**
Blackburn with Darwen UA	**2,143**	475	22	48	2	523	24
Greater Manchester (Met County)	**31,209**	3,263	10	2,017	6	5,280	17
Bolton	**3,329**	395	12	128	4	523	16
Manchester	**5,956**	927	16	876	15	1,803	30
Oldham	**3,093**	768	25	79	3	847	27
Rochdale	**2,697**	449	17	80	3	529	20
Lancashire	**12,463**	941	8	403	3	1,344	11
Burnley	**1,092**	164	15	28	3	192	18
Pendle	**1,114**	250	22	26	2	276	25
Preston	**1,691**	208	12	67	4	275	16
YORKSHIRE AND THE HUMBER	**57,923**	**4,785**	**8**	**2,716**	**5**	**7,501**	**13**
South Yorkshire (Met County)	**14,472**	710	5	763	5	1,473	10
Sheffield	**5,894**	467	8	479	8	946	16
West Yorkshire (Met County)	**26,738**	3,786	14	1,168	4	4,954	19
Bradford	**7,495**	2,008	27	249	3	2,257	30
Kirklees	**5,255**	769	15	162	3	931	18
EAST MIDLANDS	**46,916**	**2,785**	**6**	**2,520**	**5**	**5,305**	**11**
Derby UA	**2,935**	319	11	180	6	499	17
Leicester UA	**4,380**	1,092	25	550	13	1,642	37
Nottingham UA	**3,540**	367	10	261	7	628	18
Leicestershire	**6,556**	280	4	230	4	510	8
Oadby and Wigston	**495**	49	10	31	6	80	16
Northamptonshire	**7,840**	360	5	563	7	923	12
Northampton	**2,651**	192	7	281	11	473	18
WEST MIDLANDS	**63,694**	**7,111**	**11**	**3,277**	**5**	**10,388**	**16**
West Midlands (Met County)	**34,862**	6,152	18	2,200	6	8,352	24
Birmingham	**15,324**	3,928	26	1,217	8	5,145	34
Coventry	**3,801**	464	12	341	9	805	21
Sandwell	**3,959**	665	17	174	4	839	21
Walsall	**3,337**	449	13	106	3	555	17
Wolverhampton	**3,055**	353	12	136	4	489	16
EAST	**62,711**	**3,637**	**6**	**5,228**	**8**	**8,865**	**14**
Luton UA	**3,085**	992	32	322	10	1,314	43
Peterborough UA	**2,213**	292	13	222	10	514	23
Bedfordshire	**4,602**	327	7	353	8	680	15
Bedford	**1,804**	240	13	166	9	406	23
Cambridgeshire	**6,250**	227	4	736	12	963	15
Cambridge	**1,072**	84	8	253	24	337	31
Hertfordshire	**12,865**	866	7	1,189	9	2,055	16
Hertsmere	**1,086**	85	8	112	10	197	18
St Albans	**1,805**	178	10	225	12	403	22
Three Rivers	**966**	58	6	97	10	155	16
Watford	**1,105**	154	14	136	12	290	26
Welwyn Hatfield	**1,086**	73	7	133	12	206	19
Suffolk	**7,386**	160	2	698	9	858	12
Forest Heath	**781**	13	2	330	42	343	44

1 This table excludes births to mothers whose usual residence was outside of England and Wales.
2 Counties which include county districts where the proportion of non-UK born mothers is more than 15 per cent have been included to allow for comparison.
3 See section 2.7.

Table 9.2 - *continued*

Area of usual residence of mother	All live births	Birthplace of mother outside United Kingdom					
		New Commonwealth		Rest of the World		All outside United Kingdom	
		Number	Percentage	Number	Percentage	Number	Percentage
LONDON	**110,437**	**23,882**	**22**	**28,408**	**26**	**52,290**	**47**
Inner London	**47,848**	11,688	24	14,741	31	26,429	55
Camden	**2,944**	563	19	1,235	42	1,798	61
City of London	**62**	14	23	22	35	36	58
Hackney	**4,261**	998	23	1,270	30	2,268	53
Hammersmith and Fulham	**2,559**	214	8	996	39	1,210	47
Haringey	**3,890**	708	18	1,513	39	2,221	57
Islington	**2,671**	350	13	889	33	1,239	46
Kensington and Chelsea	**2,235**	141	6	1,304	58	1,445	65
Lambeth	**4,787**	1,026	21	1,413	30	2,439	51
Lewisham	**3,932**	895	23	810	21	1,705	43
Newham	**5,102**	2,328	46	1,122	22	3,450	68
Southwark	**4,342**	1,313	30	1,014	23	2,327	54
Tower Hamlets	**3,940**	2,108	54	580	15	2,688	68
Wandsworth	**4,359**	642	15	1,094	25	1,736	40
Westminster, City of	**2,764**	388	14	1,479	54	1,867	68
Outer London	**62,589**	12,194	19	13,667	22	25,861	41
Barking and Dagenham	**2,594**	508	20	430	17	938	36
Barnet	**4,334**	678	16	1,378	32	2,056	47
Bexley	**2,640**	223	8	209	8	432	16
Brent	**4,376**	1,302	30	1,547	35	2,849	65
Bromley	**3,651**	281	8	447	12	728	20
Croydon	**4,591**	1,035	23	669	15	1,704	37
Ealing	**4,479**	1,094	24	1,455	32	2,549	57
Enfield	**4,087**	774	19	1,195	29	1,969	48
Greenwich	**3,446**	663	19	678	20	1,341	39
Harrow	**2,848**	885	31	656	23	1,541	54
Hillingdon	**3,334**	597	18	641	19	1,238	37
Hounslow	**3,306**	744	23	885	27	1,629	49
Kingston upon Thames	**1,859**	226	12	406	22	632	34
Merton	**2,737**	615	22	603	22	1,218	45
Redbridge	**3,376**	1,067	32	533	16	1,600	47
Richmond upon Thames	**2,548**	202	8	580	23	782	31
Sutton	**2,247**	192	9	293	13	485	22
Waltham Forest	**3,770**	978	26	922	24	1,900	50
SOUTH EAST	**91,842**	**5,407**	**6**	**8,243**	**9**	**13,650**	**15**
Bracknell Forest UA	**1,425**	70	5	171	12	241	17
Brighton and Hove UA	**3,043**	119	4	393	13	512	17
Milton Keynes UA	**3,140**	366	12	317	10	683	22
Reading UA	**2,013**	319	16	250	12	569	28
Slough UA	**1,983**	571	29	333	17	904	46
Southampton UA	**2,556**	181	7	229	9	410	16
Windsor and Maidenhead UA	**1,674**	146	9	259	15	405	24
Wokingham UA	**1,653**	107	6	158	10	265	16
Buckinghamshire	**5,579**	608	11	471	8	1,079	19
Chiltern	**957**	79	8	93	10	172	18
South Bucks	**634**	49	8	82	13	131	21
Wycombe	**2,012**	315	16	175	9	490	24
Oxfordshire	**7,297**	443	6	839	11	1,282	18
Oxford	**1,652**	205	12	352	21	557	34
Surrey	**12,304**	729	6	1,397	11	2,126	17
Elmbridge	**1,520**	106	7	233	15	339	22
Epsom and Ewell	**790**	56	7	106	13	162	21
Guildford	**1,436**	61	4	195	14	256	18
Runnymede	**873**	43	5	113	13	156	18
Spelthorne	**1,059**	59	6	116	11	175	17
Surrey Heath	**922**	48	5	106	11	154	17
Woking	**1,184**	157	13	141	12	298	25
West Sussex	**7,909**	397	5	628	8	1,025	13
Crawley	**1,326**	200	15	128	10	328	25
SOUTH WEST	**51,476**	**1,432**	**3**	**3,059**	**6**	**4,491**	**9**
Bristol, City of UA	**5,023**	362	7	553	11	915	18
WALES	**31,400**	**759**	**2**	**1,286**	**4**	**2,045**	**7**
Cardiff	**3,708**	302	8	340	9	642	17

Table 9.3   Live births: country of birth of mother and of father[4], 2003   **England and Wales**

Country of birth of father	Country of birth of mother					New Commonwealth				
	Total	United Kingdom[1]	Total outside United Kingdom	Irish Republic	Australia Canada and New Zealand	Total	India	Pakistan	Bangladesh	East Africa
Total	621,469	506,076	115,360	3,734	4,132	55,777	7,995	15,107	8,898	3,985
United Kingdom[1]	466,602	429,594	36,994	2,468	2,742	15,625	2,701	5,459	687	1,299
Total outside United Kingdom	109,972	37,498	72,470	1,007	1,304	37,621	5,252	9,535	8,171	2,309
Irish Republic	3,277	2,247	1,030	669	35	74	7	2	2	11
Australia, Canada and New Zealand	3,689	2,323	1,366	46	837	117	3	2	-	17
New Commonwealth	59,286	20,847	38,437	117	161	35,971	5,103	9,409	8,151	1,995
India	7,112	2,285	4,827	12	9	4,614	4,246	55	16	214
Pakistan	16,948	7,283	9,665	14	10	9,393	62	9,196	24	75
Bangladesh	9,515	1,348	8,167	3	1	8,126	16	11	8,094	2
East Africa	4,366	1,555	2,811	15	34	2,376	608	127	6	1,516
Southern Africa	2,800	1,169	1,631	19	30	1,298	19	6	1	36
Rest of Africa	7,764	2,151	5,612	28	18	5,071	60	4	2	109
Caribbean	5,286	3,233	2,052	14	18	1,824	9	2	5	22
Far East[2]	1,319	685	634	4	17	467	24	4	1	7
Mediterranean[3]	1,249	963	286	5	8	142	2	1	-	4
Rest of New Commonwealth	2,927	175	2,752	3	16	2,660	57	3	2	10
Other European Union	9,253	5,255	3,997	54	132	355	22	29	7	52
Rest of Europe	7,336	1,537	5,799	21	26	115	5	7	2	15
United States of America	2,694	1,133	1,561	23	36	69	10	2	-	8
Rest of the World	24,437	4,156	20,280	77	77	920	102	84	9	211
Not stated	44,895	38,984	5,896	259	86	2,531	42	113	40	377

Country of birth of father	Country of birth of mother						Other European Union	Rest of Europe	Rest of the World	United States of America	Not Stated
	New Commonwealth - continued										
	Southern Africa	Rest of Africa	Caribbean	Far East[2]	Mediterranean[3]	Rest of New Commonwealth					
Total	3,149	7,456	3,984	1,433	945	2,825	12,152	9,475	3,037	27,053	33
United Kingdom[1]	1,352	1,387	1,136	735	690	179	7,155	2,586	1,439	4,979	14
Total outside United Kingdom	1,636	5,172	2,026	682	214	2,624	4,453	6,500	1,535	20,050	4
Irish Republic	21	9	10	8	3	1	104	32	33	83	-
Australia, Canada and New Zealand	36	8	13	21	8	9	142	56	62	106	-
New Commonwealth	1,329	4,890	1,882	485	136	2,591	628	358	93	1,109	2
India	5	43	9	20	1	5	57	35	11	89	-
Pakistan	4	14	9	3	5	1	72	34	11	131	-
Bangladesh	1	-	-	1	-	1	8	3	1	25	-
East Africa	55	26	19	7	3	9	69	37	7	273	-
Southern Africa	1,202	20	6	6	2	-	88	38	19	139	-
Rest of Africa	40	4,716	123	8	2	7	191	81	13	210	1
Caribbean	8	60	1,707	5	2	4	55	30	14	97	1
Far East[2]	5	2	1	417	3	3	27	9	9	101	-
Mediterranean[3]	6	1	5	4	118	1	29	76	6	20	-
Rest of New Commonwealth	3	8	3	14	-	2,560	32	15	2	24	-
Other European Union	59	86	37	41	17	5	2,481	256	132	587	1
Rest of Europe	30	11	1	3	36	5	161	5,311	30	135	-
United States of America	12	9	17	7	-	4	145	53	1,076	159	-
Rest of the World	149	159	66	117	14	9	792	434	109	17,871	1
Not stated	161	897	822	16	41	22	544	389	63	2,024	15

1 Including Isle of Man and Channel Islands.
2 Brunei, Malaysia and Singapore.
3 Cyprus,Gibraltar and Malta.
4 See section 2.7.

Table 9.4   Live births: age of mother and country of birth of mother[5], 2003

England and Wales

Country of birth of mother	Age of mother at birth							
	All ages	Under 20	20-24	25-29	30-34	35-39	40-44	45 and over
Total[1]	**621,469**	**44,236**	**116,622**	**156,931**	**187,214**	**97,386**	**18,205**	**875**
United Kingdom[2]	**506,076**	40,289	95,628	122,092	152,953	80,075	14,453	586
Total outside United Kingdom	**115,360**	**3,944**	**20,991**	**34,828**	**34,255**	**17,304**	**3,750**	**288**
New Commonwealth	**55,777**	**1,789**	**12,047**	**18,233**	**14,884**	**6,998**	**1,689**	**137**
India	**7,995**	137	1,505	3,129	2,165	856	184	19
Pakistan	**15,107**	512	4,409	5,229	3,319	1,299	309	30
Bangladesh	**8,898**	362	3,168	3,171	1,563	519	96	19
East Africa	**3,985**	99	358	956	1,429	893	237	13
Southern Africa	**3,149**	83	318	1,052	1,261	368	62	5
Rest of Africa	**7,456**	183	814	2,194	2,399	1,484	352	30
Caribbean	**3,984**	325	895	983	952	577	239	13
Far East[3]	**1,433**	6	66	277	572	433	77	2
Mediterranean[4]	**945**	37	101	215	330	204	53	5
Rest of New Commonwealth	**2,825**	45	413	1,027	894	365	80	1
Rest of the World	**59,583**	2,155	8,944	16,595	19,371	10,306	2,061	151

1 Includes 33 births to women whose country of birth was not stated.
2 Including Isle of Man and Channel Islands.
3 Brunei, Malaysia and Singapore.
4 Cyprus, Gibraltar and Malta.
5 See section 2.7.

Table 9.5   Total fertility rates: country of birth of mother, 1991 and 2001

England and Wales

Country of birth of mother	1991	2001
Total	**1.8**	**1.6**
United Kingdom[1]	1.8	1.6
Total outside UK	**2.3**	**2.2**
New Commonwealth	**2.8**	**2.8**
India	2.5	2.3
Pakistan	4.8	4.7
Bangladesh	5.3	3.9
East Africa	1.9	1.6
Rest of Africa[2]	2.7	2.0
Rest of New Commonwealth[3]	1.9	2.2
Rest of the World	1.9	1.8

Note: See sections 1.1, 2.7 and 2.11.
1 Including Isle of Man and Channel Islands.
2 Includes countries listed under Southern Africa and Rest of Africa in section 2.7.
3 Includes countries listed under Far East, Mediterranean, Caribbean and Rest of New Commonwealth in section 2.7.

Table 9.6 Live births (numbers and percentages): occurrence within/outside marriage, number of previous live-born children and country of birth of mother, 1993, 2001-2003

Country of birth of mother	Year	All live births (= 100%) (numbers)	All live births within marriage (percentages)	Number of previous live-born children within marriage						Births outside marriage (percentages)
				0	1	2 (percentages)	3	4	5 and over	
Total¹	**1993**	**673,467**	**67.8**	**26.5**	**25.2**	**10.7**	**3.5**	**1.2**	**0.9**	**32.2**
	2001	**594,634**	**60.0**	**24.2**	**22.2**	**8.8**	**3.0**	**1.0**	**0.8**	**40.0**
	2002	**596,122**	**59.4**	**24.4**	**21.9**	**8.4**	**2.9**	**1.0**	**0.8**	**40.6**
	2003	**621,469**	**58.6**	**24.3**	**21.4**	**8.4**	**2.8**	**1.0**	**0.7**	**41.4**
United Kingdom²	1993	592,056	65.5	25.8	24.9	10.3	3.1	0.9	0.5	34.5
	2001	496,713	55.6	22.3	21.5	8.0	2.5	0.8	0.6	44.4
	2002	490,711	54.6	22.2	21.1	7.6	2.4	0.8	0.5	45.4
	2003	506,076	53.6	22.0	20.5	7.5	2.3	0.7	0.5	46.4
Irish Republic	1993	5,149	71.0	29.5	23.8	9.9	4.0	1.6	2.3	29.0
	2001	3,843	67.6	27.7	23.1	10.6	3.4	1.2	1.6	32.4
	2002	3,708	67.4	28.2	23.3	9.5	3.1	1.4	2.0	32.6
	2003	3,734	67.9	29.1	22.5	10.1	3.2	1.3	1.7	32.1
New Commonwealth	**1993**	**47,793**	**88.1**	**28.5**	**27.3**	**15.6**	**8.0**	**4.1**	**4.7**	**11.9**
	2001	**50,047**	**87.0**	**33.4**	**26.4**	**14.7**	**7.1**	**3.1**	**2.4**	**13.0**
	2002	**54,037**	**86.1**	**34.0**	**26.2**	**14.0**	**6.8**	**2.8**	**2.2**	**13.9**
	2003	**55,777**	**85.6**	**33.3**	**26.5**	**14.4**	**6.8**	**2.8**	**1.9**	**14.4**
India	1993	7,316	97.5	35.0	33.2	18.0	7.2	2.4	1.7	2.5
	2001	6,598	97.9	45.1	33.9	12.9	4.0	1.5	0.5	2.1
	2002	7,222	97.8	47.9	33.1	11.3	3.6	1.2	0.7	2.2
	2003	7,995	97.6	49.2	32.8	11.0	3.2	0.9	0.6	2.4
Pakistan	1993	13,023	98.7	25.8	25.2	17.7	12.7	7.9	9.4	1.3
	2001	14,588	98.4	33.5	25.6	18.5	11.6	5.3	3.9	1.6
	2002	15,357	98.4	33.5	26.6	18.4	11.2	4.9	3.8	1.6
	2003	15,107	98.2	29.6	28.0	20.2	11.7	5.4	3.3	1.8
Bangladesh	1993	5,854	98.9	29.3	23.2	15.3	10.4	7.7	12.9	1.1
	2001	8,164	98.9	31.3	27.2	19.2	10.2	5.6	5.4	1.1
	2002	8,486	98.7	31.3	27.6	19.0	10.9	5.2	4.7	1.3
	2003	8,898	98.6	30.3	27.4	20.2	11.2	5.1	4.3	1.4
East Africa	1993	5,698	90.0	33.0	33.9	15.5	5.4	1.6	0.6	10.0
	2001	3,745	80.1	31.4	29.0	13.2	4.6	1.2	0.7	19.9
	2002	3,724	77.8	30.5	28.3	13.2	3.9	1.1	0.7	22.2
	2003	3,985	75.3	31.5	26.8	11.8	3.6	1.2	0.5	24.7
Southern Africa	1993	:	:	:	:	:	:	:	:	:
	2001	2,236	79.7	43.1	25.3	8.5	2.0	0.4	0.4	20.3
	2002	2,654	79.7	44.8	26.0	6.3	2.0	0.3	0.4	20.3
	2003	3,149	78.3	44.6	25.4	6.4	1.3	0.3	0.3	21.7
Rest of Africa	1993	5,511	60.8	19.8	21.3	12.3	5.1	1.6	0.7	39.2
	2001	6,995	65.3	23.9	22.2	12.3	4.7	1.5	0.7	34.7
	2002	8,073	64.9	26.5	20.4	11.6	4.3	1.4	0.8	35.1
	2003	7,456	64.1	26.1	19.5	12.0	4.8	1.1	0.6	35.9
Caribbean	1993	3,124	51.9	15.7	17.1	11.9	4.4	1.5	1.3	48.1
	2001	3,085	42.2	17.0	13.7	6.6	2.9	1.3	0.7	57.8
	2002	3,598	38.0	15.3	12.2	6.2	2.4	1.1	0.8	62.0
	2003	3,984	37.4	15.7	12.7	5.3	2.5	0.8	0.5	62.6
Far East³	1993	3,710	85.9	34.8	31.2	13.8	4.6	0.9	0.6	14.1
	2001	1,365	85.1	38.7	28.5	11.0	4.7	1.0	1.2	14.9
	2002	1,351	86.2	38.2	31.2	11.0	3.8	1.2	1.0	13.8
	2003	1,433	88.3	41.9	29.4	11.0	4.4	0.9	0.7	11.7
Mediterranean⁴	1993	1,994	80.7	28.0	31.1	15.3	4.7	1.0	0.5	19.3
	2001	1,099	75.0	28.9	27.4	13.2	4.1	0.8	0.5	25.0
	2002	1,065	74.0	28.6	28.0	11.4	4.6	0.8	0.6	26.0
	2003	945	70.2	28.0	26.6	10.8	3.2	1.1	0.5	29.8
Rest of New Commonwealth	1993	1,563	90.5	41.8	34.7	10.7	2.6	0.4	0.3	9.5
	2001	2,172	93.4	50.4	31.7	9.7	1.2	0.4	0.0	6.6
	2002	2,507	94.7	50.4	32.3	9.6	2.0	0.3	0.1	5.3
	2003	2,825	94.4	48.1	34.1	10.1	1.8	0.3	0.1	5.6
Rest of the World	1993	28,425	82.3	36.6	27.5	11.0	4.0	1.6	1.6	17.7
	2001	44,031	77.6	35.1	25.5	10.3	3.7	1.6	1.5	22.4
	2002	47,666	77.5	35.6	24.6	10.2	3.9	1.5	1.6	22.5
	2003	55,882	76.3	35.5	24.2	9.6	3.9	1.5	1.6	23.7

Note: See section 2.7.
1 Includes births to women whose country of birth was not stated.
2 Including Isle of Man and Channel Islands.
3 Brunei, Malaysia and Singapore, and Hong Kong before 1997.
4 Cyprus, Gibraltar and Malta.

Table 10.1 Age-specific fertility rates[1]: age and year of birth of woman, 1920-1988

Year of birth of woman/female birth cohort	Age of woman - completed years														
	15	16	17	18	19	20	21	22	23	24	25	26	27	28	29
1920	0	3	10	23	40	59	84	104	117	131	119	159	165	136	123
1925	0	2	9	22	43	67	101	147	157	156	147	141	134	131	118
1926	1	3	9	23	44	72	124	142	151	150	148	148	142	133	121
1927	1	3	10	24	45	86	124	138	144	149	148	152	142	134	128
1928	1	3	11	24	54	88	123	134	146	152	155	150	143	143	135
1929	1	4	10	27	59	90	117	132	146	158	156	151	148	148	137
1930	1	3	11	30	60	87	117	138	155	162	161	165	160	154	141
1931	1	3	12	32	58	85	119	143	156	162	169	170	162	153	146
1932	1	3	14	31	57	87	123	142	154	171	175	173	160	156	147
1933	1	4	13	31	58	90	125	146	167	180	183	178	172	165	151
1934	1	3	12	32	61	92	128	157	175	188	190	189	179	173	155
1935	1	3	13	33	61	94	137	162	180	190	198	193	181	173	158
1936	1	3	13	35	63	103	141	167	179	191	201	196	184	174	151
1937	1	3	14	34	70	107	147	168	184	198	206	199	188	167	142
1938	1	3	15	40	75	112	150	174	194	204	210	202	181	162	135
1939	1	4	18	44	79	116	155	180	200	208	213	193	174	154	131
1940	1	5	20	48	83	123	162	189	204	214	208	191	169	152	127
1941	1	5	22	52	91	127	169	193	209	210	201	183	162	144	123
1942	1	6	24	54	91	132	169	192	202	202	193	178	159	139	121
1943	1	7	27	57	95	132	167	187	197	191	187	172	154	139	112
1944	1	8	31	63	98	135	166	183	188	189	180	168	154	128	105
1945	2	11	35	66	103	136	166	176	181	181	178	168	145	121	101
1946	2	12	36	67	104	136	161	172	175	179	178	159	139	119	100
1947	2	12	36	66	99	128	151	158	165	169	159	145	130	111	97
1948	3	13	38	67	102	127	147	161	167	160	153	142	127	114	100
1949	2	13	38	71	101	124	144	155	152	148	144	134	125	113	106
1950	3	14	41	73	101	124	142	141	142	141	137	131	123	118	114
1951	3	15	43	75	102	125	129	130	133	130	131	126	126	127	115
1952	3	16	45	77	104	115	121	126	125	128	129	132	136	129	114
1953	3	17	46	78	97	107	115	118	121	122	133	141	138	125	114
1954	4	17	47	73	89	99	105	111	114	127	139	142	133	122	113
1955	4	18	47	68	83	92	101	107	119	134	144	135	129	123	114
1956	4	19	42	62	76	86	95	110	126	136	138	133	129	124	117
1957	4	18	41	57	70	81	98	115	128	131	133	134	130	126	116
1958	4	16	35	50	63	81	100	116	119	127	132	133	131	123	118
1959	4	13	30	46	63	83	102	107	116	124	129	132	128	125	119
1960	3	12	28	45	66	85	96	105	114	123	129	128	130	125	117
1961	3	11	27	47	66	78	90	102	110	119	124	129	127	122	120
1962	3	11	28	47	61	74	86	98	108	115	123	126	124	125	117
1963	3	11	27	42	58	71	83	96	105	114	122	122	127	122	117
1964	3	11	24	41	55	68	83	93	104	114	117	123	123	121	116
1965	3	10	24	39	55	69	81	93	104	109	119	120	120	119	115
1966	2	10	24	40	58	70	82	94	101	109	115	119	117	116	113
1967	2	10	25	43	59	71	85	92	100	107	111	114	115	112	111
1968	3	11	27	45	61	74	83	91	99	104	107	111	111	112	111
1969	3	12	28	46	63	72	82	89	94	99	103	106	109	110	109
1970	3	12	28	48	61	72	80	86	91	95	99	104	106	107	106
1971	3	12	30	46	62	71	76	81	85	89	97	101	104	104	103
1972	3	13	29	47	61	68	73	77	81	87	92	97	99	99	99
1973	3	13	31	47	59	66	70	75	81	86	90	94	94	95	99
1974	4	13	30	45	57	65	70	77	80	85	87	90	93	96	106
1975	4	14	30	44	55	63	70	75	78	81	84	87	92	102	
1976	4	13	30	43	56	67	71	75	76	78	81	87	98		
1977	4	14	28	44	58	66	69	73	73	75	81	91			
1978	4	13	28	46	58	65	69	70	72	75	84				
1979	4	13	31	46	58	65	67	69	72	79					
1980	4	14	31	46	59	63	66	70	75						
1981	4	14	31	46	57	63	66	70							
1982	4	14	30	45	56	62	68								
1983	4	13	29	43	54	63									
1984	4	12	27	41	54										
1985	4	11	26	41											
1986	4	11	26												
1987	3	11													
1988	3														

Notes: The age-specific fertility rates refer to 'all live births per 1,000 women' at the age shown.
Live births to women aged under 15 are not included in the calculation of the rate for age 15.
1 See sections 2.11 and 3.9.
2 Includes births at ages 45 and over, achieved up to the end of 2003 by women born in 1958 and earlier years.

30	31	32	33	34	35	36	37	38	39	40	41	42	43	44	45²	Year of birth of woman/female birth cohort
109	93	87	76	67	58	52	44	38	32	25	17	13	8	5	4	1920
108	97	90	78	72	65	56	47	40	32	23	15	11	6	3	3	1925
113	100	94	80	76	67	58	48	43	31	22	15	10	5	3	3	1926
119	103	95	85	77	67	57	49	40	29	21	13	9	5	3	2	1927
123	104	101	86	78	67	59	46	38	28	20	12	8	5	2	2	1928
125	109	102	86	79	68	55	43	36	26	18	11	7	4	2	2	1929
136	117	109	91	83	68	54	43	34	24	17	11	7	4	2	2	1930
134	117	107	91	77	63	51	39	30	21	16	10	6	3	2	2	1931
132	116	111	88	72	60	47	36	28	20	13	8	5	3	2	2	1932
137	117	105	85	70	57	44	34	26	18	11	7	4	2	1	2	1933
140	112	97	79	67	53	42	32	22	15	10	6	4	2	1	2	1934
134	104	92	73	62	50	39	28	19	13	9	6	4	2	2	2	1935
127	101	88	70	57	47	34	23	17	12	8	5	4	2	2	2	1936
121	97	83	66	54	41	29	21	16	11	7	5	4	2	1	2	1937
117	94	78	64	47	35	25	19	14	11	8	5	4	2	1	2	1938
111	90	75	56	41	31	24	18	14	10	8	6	4	2	1	2	1939
108	88	67	50	38	29	23	17	14	11	8	7	4	2	1	2	1940
107	80	60	46	36	28	22	18	15	12	8	6	4	2	2	2	1941
96	72	57	43	34	28	23	20	16	12	8	6	4	2	1	2	1942
89	68	54	43	35	30	25	21	15	11	8	6	4	2	1	2	1943
85	66	53	43	37	32	27	21	15	11	8	6	3	2	1	2	1944
83	66	56	47	42	34	26	21	16	12	9	6	4	2	1	2	1945
85	69	60	54	44	34	28	22	17	13	9	6	4	2	1	2	1946
83	73	64	53	41	34	27	22	16	12	9	6	4	2	1	1	1947
92	82	69	56	45	36	29	23	18	13	9	6	4	3	1	1	1948
99	85	69	55	46	37	31	24	19	14	10	7	4	3	1	1	1949
103	85	68	56	47	39	31	26	20	15	11	7	5	3	2	2	1950
100	83	70	58	49	42	34	27	21	16	11	8	5	3	2	2	1951
100	86	72	62	51	44	35	28	22	17	12	8	5	3	2	2	1952
102	88	76	64	54	45	37	30	23	18	13	8	5	3	2	2	1953
103	90	77	67	57	47	39	30	24	18	12	9	6	3	2	2	1954
106	91	81	68	58	49	40	32	24	19	14	9	6	4	2	2	1955
106	94	82	68	61	50	41	33	25	20	14	10	6	4	2	2	1956
109	94	82	72	62	52	42	34	26	20	15	10	6	4	2	3	1957
108	96	86	73	63	52	44	35	27	21	15	10	7	4	2	2	1958
108	99	86	74	63	55	44	35	29	22	16	11	7	4	2		1959
113	98	88	75	66	55	47	38	30	23	16	11	7	5			1960
111	99	87	76	66	56	48	38	29	23	17	11	8				1961
110	99	90	77	68	59	49	38	30	23	17	12					1962
109	101	89	79	69	60	49	40	31	24	19						1963
111	99	91	81	71	61	51	41	32	27							1964
109	102	92	82	72	61	51	42	35								1965
108	100	93	82	71	62	53	46									1966
107	100	91	80	71	63	56										1967
108	99	90	81	74	69											1968
105	97	91	83	79												1969
102	98	93	89													1970
101	99	99														1971
101	103															1972
105																1973
																1974
																1975
																1976
																1977
																1978
																1979
																1980
																1981
																1982
																1983
																1984
																1985
																1986
																1987
																1988

Table 10.2 Average number of live-born children[1]: age and year of birth of woman, 1920-1988

Year of birth of woman/female birth cohort	Age of woman - completed years														
	15	16	17	18	19	20	21	22	23	24	25	26	27	28	29
1920	0.00	0.00	0.01	0.04	0.08	0.13	0.22	0.32	0.44	0.57	0.69	0.85	1.01	1.15	1.27
1925	0.00	0.00	0.01	0.03	0.08	0.14	0.24	0.39	0.55	0.70	0.85	0.99	1.13	1.26	1.37
1926	0.00	0.00	0.01	0.04	0.08	0.15	0.28	0.42	0.57	0.72	0.87	1.01	1.16	1.29	1.41
1927	0.00	0.00	0.01	0.04	0.08	0.17	0.29	0.43	0.57	0.72	0.87	1.02	1.17	1.30	1.43
1928	0.00	0.00	0.02	0.04	0.09	0.18	0.30	0.44	0.58	0.74	0.89	1.04	1.18	1.33	1.46
1929	0.00	0.00	0.01	0.04	0.10	0.19	0.31	0.44	0.59	0.75	0.90	1.05	1.20	1.35	1.49
1930	0.00	0.00	0.01	0.04	0.10	0.19	0.31	0.45	0.60	0.76	0.93	1.09	1.25	1.41	1.55
1931	0.00	0.00	0.02	0.05	0.11	0.19	0.31	0.45	0.61	0.77	0.94	1.11	1.27	1.42	1.57
1932	0.00	0.00	0.02	0.05	0.11	0.19	0.32	0.46	0.61	0.78	0.96	1.13	1.29	1.45	1.59
1933	0.00	0.00	0.02	0.05	0.11	0.20	0.32	0.47	0.63	0.81	1.00	1.18	1.35	1.51	1.66
1934	0.00	0.00	0.02	0.05	0.11	0.20	0.33	0.49	0.66	0.85	1.04	1.23	1.41	1.58	1.73
1935	0.00	0.00	0.02	0.05	0.11	0.20	0.34	0.50	0.68	0.87	1.07	1.26	1.45	1.62	1.78
1936	0.00	0.00	0.02	0.05	0.12	0.22	0.36	0.53	0.71	0.90	1.10	1.29	1.48	1.65	1.80
1937	0.00	0.00	0.02	0.05	0.12	0.23	0.38	0.55	0.73	0.93	1.13	1.33	1.52	1.69	1.83
1938	0.00	0.00	0.02	0.06	0.13	0.25	0.40	0.57	0.76	0.97	1.18	1.38	1.56	1.72	1.86
1939	0.00	0.00	0.02	0.07	0.14	0.26	0.42	0.60	0.80	1.00	1.22	1.41	1.58	1.74	1.87
1940	0.00	0.01	0.03	0.07	0.16	0.28	0.44	0.63	0.83	1.05	1.26	1.45	1.61	1.77	1.89
1941	0.00	0.01	0.03	0.08	0.17	0.30	0.47	0.66	0.87	1.08	1.28	1.46	1.63	1.77	1.89
1942	0.00	0.01	0.03	0.08	0.18	0.31	0.48	0.67	0.87	1.07	1.27	1.44	1.60	1.74	1.86
1943	0.00	0.01	0.04	0.09	0.19	0.32	0.49	0.67	0.87	1.06	1.25	1.42	1.57	1.71	1.83
1944	0.00	0.01	0.04	0.10	0.20	0.34	0.50	0.69	0.87	1.06	1.24	1.41	1.56	1.69	1.80
1945	0.00	0.01	0.05	0.11	0.22	0.35	0.52	0.69	0.88	1.06	1.23	1.40	1.55	1.67	1.77
1946	0.00	0.01	0.05	0.12	0.22	0.36	0.52	0.69	0.86	1.04	1.22	1.38	1.52	1.64	1.74
1947	0.00	0.01	0.05	0.12	0.22	0.34	0.49	0.65	0.82	0.99	1.14	1.29	1.42	1.53	1.63
1948	0.00	0.02	0.05	0.12	0.22	0.35	0.50	0.66	0.83	0.99	1.14	1.28	1.41	1.52	1.62
1949	0.00	0.02	0.05	0.13	0.23	0.35	0.50	0.65	0.80	0.95	1.09	1.23	1.35	1.47	1.57
1950	0.00	0.02	0.06	0.13	0.23	0.36	0.50	0.64	0.78	0.92	1.06	1.19	1.31	1.43	1.54
1951	0.00	0.02	0.06	0.14	0.24	0.36	0.49	0.62	0.76	0.89	1.02	1.14	1.27	1.40	1.51
1952	0.00	0.02	0.07	0.14	0.25	0.36	0.48	0.61	0.73	0.86	0.99	1.12	1.26	1.39	1.50
1953	0.00	0.02	0.07	0.14	0.24	0.35	0.46	0.58	0.70	0.82	0.96	1.10	1.24	1.36	1.48
1954	0.00	0.02	0.07	0.14	0.23	0.33	0.44	0.55	0.66	0.79	0.93	1.07	1.20	1.32	1.44
1955	0.00	0.02	0.07	0.14	0.22	0.31	0.41	0.52	0.64	0.77	0.92	1.05	1.18	1.31	1.42
1956	0.00	0.02	0.07	0.13	0.20	0.29	0.39	0.50	0.62	0.76	0.90	1.03	1.16	1.28	1.40
1957	0.00	0.02	0.06	0.12	0.19	0.27	0.37	0.49	0.61	0.74	0.88	1.01	1.14	1.27	1.38
1958	0.00	0.02	0.06	0.11	0.17	0.25	0.35	0.47	0.58	0.71	0.84	0.98	1.11	1.23	1.35
1959	0.00	0.02	0.05	0.09	0.16	0.24	0.34	0.45	0.56	0.69	0.82	0.95	1.08	1.20	1.32
1960	0.00	0.02	0.04	0.09	0.15	0.24	0.34	0.44	0.55	0.68	0.81	0.93	1.06	1.19	1.31
1961	0.00	0.02	0.04	0.09	0.16	0.23	0.32	0.42	0.53	0.65	0.78	0.91	1.03	1.16	1.28
1962	0.00	0.01	0.04	0.09	0.15	0.23	0.31	0.41	0.52	0.63	0.76	0.88	1.01	1.13	1.25
1963	0.00	0.01	0.04	0.08	0.14	0.21	0.30	0.39	0.50	0.61	0.73	0.86	0.98	1.10	1.22
1964	0.00	0.01	0.04	0.08	0.13	0.20	0.29	0.38	0.48	0.60	0.71	0.84	0.96	1.08	1.20
1965	0.00	0.01	0.04	0.08	0.13	0.20	0.28	0.38	0.48	0.59	0.71	0.83	0.95	1.06	1.18
1966	0.00	0.01	0.04	0.08	0.13	0.20	0.29	0.38	0.48	0.59	0.71	0.82	0.94	1.06	1.17
1967	0.00	0.01	0.04	0.08	0.14	0.21	0.30	0.39	0.49	0.60	0.71	0.82	0.94	1.05	1.16
1968	0.00	0.01	0.04	0.09	0.15	0.22	0.30	0.40	0.49	0.60	0.71	0.82	0.93	1.04	1.15
1969	0.00	0.02	0.04	0.09	0.15	0.22	0.31	0.39	0.49	0.59	0.69	0.80	0.91	1.02	1.12
1970	0.00	0.02	0.04	0.09	0.15	0.22	0.30	0.39	0.48	0.58	0.67	0.78	0.88	0.99	1.10
1971	0.00	0.02	0.05	0.09	0.15	0.22	0.30	0.38	0.47	0.56	0.65	0.75	0.86	0.96	1.06
1972	0.00	0.02	0.05	0.09	0.15	0.22	0.30	0.37	0.45	0.54	0.63	0.73	0.83	0.93	1.03
1973	0.00	0.02	0.05	0.09	0.15	0.22	0.29	0.36	0.44	0.53	0.62	0.71	0.81	0.90	1.00
1974	0.00	0.02	0.05	0.09	0.15	0.22	0.28	0.36	0.44	0.53	0.61	0.70	0.80	0.89	1.00
1975	0.00	0.02	0.05	0.09	0.15	0.21	0.28	0.35	0.43	0.51	0.60	0.69	0.78	0.88	
1976	0.00	0.02	0.05	0.09	0.15	0.21	0.28	0.36	0.44	0.51	0.60	0.68	0.78		
1977	0.00	0.02	0.05	0.09	0.15	0.22	0.28	0.36	0.43	0.51	0.59	0.68			
1978	0.00	0.02	0.05	0.09	0.15	0.21	0.28	0.35	0.42	0.50	0.58				
1979	0.00	0.02	0.05	0.09	0.15	0.22	0.28	0.35	0.43	0.51					
1980	0.00	0.02	0.05	0.10	0.15	0.22	0.28	0.35	0.43						
1981	0.01	0.02	0.05	0.10	0.15	0.22	0.28	0.35							
1982	0.01	0.02	0.05	0.09	0.15	0.21	0.28								
1983	0.00	0.02	0.05	0.09	0.14	0.21									
1984	0.00	0.02	0.04	0.09	0.14										
1985	0.00	0.02	0.04	0.08											
1986	0.00	0.02	0.04												
1987	0.00	0.02													
1988	0.00														

1 See section 3.9.
2 Includes births at ages 45 and over, achieved up to the end of 2003 by women born in 1958 and earlier years.

England and Wales

30	31	32	33	34	35	36	37	38	39	40	41	42	43	44	45²	Year of birth of woman/female birth cohort
1.38	1.47	1.56	1.64	1.70	1.76	1.81	1.86	1.90	1.93	1.95	1.97	1.98	1.99	2.00	2.00	1920
1.48	1.58	1.67	1.75	1.82	1.88	1.94	1.99	2.03	2.06	2.08	2.10	2.11	2.11	2.12	2.12	1925
1.52	1.62	1.72	1.80	1.87	1.94	2.00	2.05	2.09	2.12	2.14	2.16	2.17	2.17	2.18	2.18	1926
1.55	1.65	1.74	1.83	1.91	1.97	2.03	2.08	2.12	2.15	2.17	2.18	2.19	2.19	2.20	2.20	1927
1.58	1.69	1.79	1.88	1.95	2.02	2.08	2.13	2.16	2.19	2.21	2.22	2.23	2.24	2.24	2.24	1928
1.61	1.72	1.82	1.91	1.99	2.05	2.11	2.15	2.19	2.21	2.23	2.24	2.25	2.25	2.26	2.26	1929
1.68	1.80	1.91	2.00	2.08	2.15	2.20	2.25	2.28	2.30	2.32	2.33	2.34	2.34	2.34	2.35	1930
1.70	1.82	1.93	2.02	2.10	2.16	2.21	2.25	2.28	2.30	2.32	2.33	2.33	2.34	2.34	2.34	1931
1.73	1.84	1.95	2.04	2.11	2.17	2.22	2.26	2.28	2.30	2.32	2.33	2.33	2.33	2.33	2.34	1932
1.80	1.92	2.02	2.11	2.18	2.23	2.28	2.31	2.34	2.36	2.37	2.37	2.38	2.38	2.38	2.39	1933
1.88	1.99	2.08	2.16	2.23	2.28	2.33	2.36	2.38	2.40	2.41	2.41	2.42	2.42	2.42	2.42	1934
1.91	2.01	2.11	2.18	2.24	2.29	2.33	2.36	2.38	2.39	2.40	2.41	2.41	2.41	2.41	2.42	1935
1.93	2.03	2.12	2.19	2.25	2.29	2.33	2.35	2.37	2.38	2.39	2.39	2.40	2.40	2.40	2.40	1936
1.95	2.05	2.13	2.20	2.25	2.29	2.32	2.34	2.35	2.37	2.37	2.38	2.38	2.38	2.39	2.39	1937
1.98	2.07	2.15	2.21	2.26	2.29	2.32	2.34	2.35	2.36	2.37	2.38	2.38	2.38	2.38	2.39	1938
1.98	2.07	2.15	2.20	2.24	2.27	2.30	2.32	2.33	2.34	2.35	2.35	2.36	2.36	2.36	2.36	1939
2.00	2.09	2.16	2.21	2.24	2.27	2.30	2.31	2.33	2.34	2.35	2.36	2.36	2.36	2.36	2.36	1940
2.00	2.08	2.14	2.19	2.22	2.25	2.27	2.29	2.31	2.32	2.33	2.33	2.33	2.34	2.34	2.34	1941
1.96	2.03	2.09	2.13	2.16	2.19	2.21	2.23	2.25	2.26	2.27	2.28	2.28	2.28	2.28	2.29	1942
1.91	1.98	2.04	2.08	2.12	2.15	2.17	2.19	2.21	2.22	2.23	2.23	2.24	2.24	2.24	2.24	1943
1.88	1.95	2.00	2.04	2.08	2.11	2.14	2.16	2.18	2.19	2.20	2.20	2.20	2.21	2.21	2.21	1944
1.85	1.92	1.97	2.02	2.06	2.10	2.12	2.14	2.16	2.17	2.18	2.19	2.19	2.19	2.19	2.19	1945
1.82	1.89	1.95	2.00	2.05	2.08	2.11	2.13	2.15	2.16	2.17	2.18	2.18	2.18	2.18	2.19	1946
1.71	1.78	1.85	1.90	1.94	1.98	2.00	2.02	2.04	2.05	2.06	2.07	2.07	2.07	2.08	2.08	1947
1.71	1.80	1.86	1.92	1.97	2.00	2.03	2.05	2.07	2.08	2.09	2.10	2.10	2.11	2.11	2.11	1948
1.67	1.76	1.82	1.88	1.93	1.96	1.99	2.02	2.04	2.05	2.06	2.07	2.07	2.07	2.07	2.08	1949
1.65	1.73	1.80	1.86	1.91	1.94	1.98	2.00	2.02	2.04	2.05	2.05	2.06	2.06	2.06	2.07	1950
1.61	1.69	1.77	1.82	1.87	1.91	1.95	1.98	2.00	2.01	2.02	2.03	2.04	2.04	2.04	2.04	1951
1.60	1.69	1.76	1.82	1.87	1.92	1.95	1.98	2.00	2.02	2.03	2.04	2.04	2.05	2.05	2.05	1952
1.58	1.67	1.74	1.81	1.86	1.91	1.94	1.97	1.99	2.01	2.03	2.03	2.04	2.04	2.04	2.05	1953
1.54	1.63	1.71	1.77	1.83	1.88	1.92	1.95	1.97	1.99	2.00	2.01	2.02	2.02	2.02	2.02	1954
1.53	1.62	1.70	1.76	1.82	1.87	1.91	1.94	1.97	1.99	2.00	2.01	2.02	2.02	2.02	2.02	1955
1.51	1.60	1.68	1.75	1.81	1.86	1.90	1.93	1.96	1.98	1.99	2.00	2.01	2.01	2.01	2.02	1956
1.49	1.59	1.67	1.74	1.80	1.85	1.89	1.93	1.95	1.97	1.99	2.00	2.01	2.01	2.01	2.01	1957
1.46	1.55	1.64	1.71	1.77	1.83	1.87	1.90	1.93	1.95	1.97	1.98	1.98	1.99	1.99	1.99	1958
1.43	1.53	1.62	1.69	1.75	1.81	1.85	1.89	1.92	1.94	1.95	1.96	1.97	1.98	1.98		1959
1.42	1.52	1.60	1.68	1.75	1.80	1.85	1.89	1.92	1.94	1.95	1.96	1.97	1.98			1960
1.39	1.49	1.57	1.65	1.71	1.77	1.82	1.86	1.89	1.91	1.93	1.94					1961
1.36	1.46	1.55	1.62	1.69	1.75	1.80	1.84	1.87	1.89	1.91	1.92					1962
1.33	1.43	1.52	1.60	1.67	1.73	1.78	1.82	1.85	1.87	1.89						1963
1.31	1.41	1.50	1.58	1.65	1.71	1.76	1.80	1.83	1.86							1964
1.29	1.39	1.48	1.56	1.64	1.70	1.75	1.79	1.83								1965
1.28	1.38	1.47	1.55	1.62	1.69	1.74	1.79									1966
1.27	1.37	1.46	1.54	1.61	1.67	1.73										1967
1.26	1.36	1.45	1.53	1.60	1.67											1968
1.23	1.33	1.42	1.50	1.58												1969
1.20	1.30	1.39	1.48													1970
																1974
1.17	1.26	1.36														1971
1.13	1.23															1972
1.11																1973
																1975
																1976
																1977
																1978
																1979
																1980
																1981
																1982
																1983
																1984
																1985
																1986
																1987
																1988

Table 10.3 Estimated average number of first live-born children[1]:
age and year of birth of woman, 1920-1988

Year of birth of woman/female birth cohort	Age of woman - completed years														
	15	16	17	18	19	20	21	22	23	24	25	26	27	28	29
1920	0.00	0.00	0.01	0.03	0.07	0.11	0.17	0.25	0.32	0.40	0.45	0.53	0.60	0.64	0.67
1925	0.00	0.00	0.01	0.03	0.07	0.12	0.20	0.31	0.40	0.48	0.54	0.60	0.64	0.68	0.71
1926	0.00	0.00	0.01	0.03	0.07	0.13	0.23	0.32	0.41	0.48	0.54	0.60	0.65	0.68	0.72
1927	0.00	0.00	0.01	0.03	0.07	0.15	0.24	0.33	0.41	0.48	0.54	0.60	0.65	0.69	0.72
1928	0.00	0.00	0.01	0.04	0.08	0.15	0.24	0.33	0.41	0.48	0.55	0.61	0.65	0.69	0.73
1929	0.00	0.00	0.01	0.04	0.09	0.16	0.24	0.32	0.40	0.48	0.55	0.61	0.65	0.70	0.73
1930	0.00	0.00	0.01	0.04	0.09	0.16	0.24	0.33	0.41	0.49	0.56	0.62	0.67	0.72	0.75
1931	0.00	0.00	0.02	0.04	0.09	0.16	0.24	0.33	0.41	0.49	0.56	0.62	0.68	0.72	0.75
1932	0.00	0.00	0.02	0.04	0.09	0.16	0.24	0.33	0.41	0.50	0.57	0.63	0.68	0.72	0.75
1933	0.00	0.00	0.02	0.04	0.09	0.16	0.25	0.34	0.43	0.51	0.59	0.65	0.70	0.74	0.77
1934	0.00	0.00	0.02	0.04	0.09	0.16	0.25	0.35	0.45	0.53	0.61	0.67	0.72	0.76	0.79
1935	0.00	0.00	0.02	0.04	0.09	0.17	0.26	0.36	0.45	0.54	0.61	0.67	0.72	0.77	0.80
1936	0.00	0.00	0.02	0.05	0.10	0.18	0.27	0.37	0.46	0.54	0.62	0.68	0.73	0.77	0.80
1937	0.00	0.00	0.02	0.05	0.10	0.18	0.28	0.38	0.47	0.55	0.63	0.69	0.73	0.77	0.80
1938	0.00	0.00	0.02	0.05	0.11	0.20	0.29	0.39	0.48	0.57	0.64	0.70	0.75	0.78	0.81
1939	0.00	0.00	0.02	0.06	0.12	0.21	0.30	0.40	0.50	0.58	0.65	0.71	0.75	0.79	0.81
1940	0.00	0.01	0.02	0.07	0.13	0.22	0.32	0.42	0.51	0.59	0.66	0.72	0.76	0.79	0.82
1941	0.00	0.01	0.03	0.07	0.14	0.23	0.33	0.43	0.52	0.61	0.67	0.73	0.77	0.80	0.82
1942	0.00	0.01	0.03	0.07	0.14	0.23	0.33	0.43	0.52	0.60	0.67	0.72	0.76	0.79	0.82
1943	0.00	0.01	0.03	0.08	0.15	0.24	0.33	0.43	0.52	0.60	0.66	0.71	0.75	0.79	0.81
1944	0.00	0.01	0.04	0.09	0.16	0.25	0.34	0.44	0.52	0.60	0.66	0.72	0.76	0.79	0.82
1945	0.00	0.01	0.04	0.10	0.17	0.26	0.35	0.44	0.53	0.60	0.66	0.72	0.76	0.80	0.82
1946	0.00	0.01	0.05	0.10	0.17	0.26	0.35	0.44	0.52	0.59	0.66	0.71	0.76	0.80	0.82
1947	0.00	0.01	0.05	0.10	0.17	0.25	0.34	0.42	0.49	0.56	0.63	0.68	0.72	0.76	0.79
1948	0.00	0.02	0.05	0.10	0.18	0.26	0.34	0.42	0.50	0.56	0.62	0.68	0.72	0.76	0.79
1949	0.00	0.02	0.05	0.11	0.18	0.26	0.34	0.41	0.48	0.54	0.60	0.66	0.70	0.74	0.77
1950	0.00	0.02	0.05	0.11	0.19	0.26	0.34	0.41	0.47	0.53	0.59	0.64	0.68	0.72	0.76
1951	0.00	0.02	0.06	0.12	0.19	0.27	0.34	0.40	0.46	0.52	0.57	0.62	0.67	0.71	0.74
1952	0.00	0.02	0.06	0.12	0.20	0.27	0.33	0.39	0.45	0.50	0.56	0.61	0.66	0.70	0.74
1953	0.00	0.02	0.06	0.13	0.19	0.26	0.32	0.38	0.43	0.48	0.54	0.60	0.65	0.69	0.72
1954	0.00	0.02	0.06	0.12	0.19	0.24	0.30	0.35	0.41	0.46	0.52	0.58	0.63	0.67	0.70
1955	0.00	0.02	0.06	0.12	0.18	0.23	0.29	0.34	0.40	0.46	0.52	0.57	0.62	0.66	0.70
1956	0.00	0.02	0.06	0.11	0.16	0.22	0.27	0.32	0.39	0.45	0.50	0.56	0.60	0.64	0.68
1957	0.00	0.02	0.06	0.11	0.15	0.20	0.26	0.32	0.38	0.44	0.49	0.54	0.59	0.63	0.67
1958	0.00	0.02	0.05	0.09	0.14	0.19	0.24	0.31	0.36	0.42	0.47	0.53	0.58	0.62	0.66
1959	0.00	0.02	0.04	0.08	0.13	0.18	0.24	0.29	0.35	0.40	0.46	0.51	0.56	0.60	0.65
1960	0.00	0.02	0.04	0.08	0.13	0.18	0.23	0.29	0.34	0.40	0.45	0.50	0.55	0.60	0.64
1961	0.00	0.01	0.04	0.08	0.13	0.18	0.23	0.28	0.33	0.38	0.44	0.49	0.54	0.58	0.63
1962	0.00	0.01	0.04	0.08	0.12	0.17	0.22	0.27	0.32	0.37	0.43	0.48	0.53	0.58	0.62
1963	0.00	0.01	0.04	0.08	0.12	0.16	0.21	0.26	0.31	0.37	0.42	0.47	0.52	0.57	0.61
1964	0.00	0.01	0.04	0.07	0.11	0.16	0.20	0.25	0.30	0.36	0.41	0.46	0.51	0.56	0.60
1965	0.00	0.01	0.04	0.07	0.11	0.15	0.20	0.25	0.30	0.35	0.40	0.46	0.51	0.55	0.59
1966	0.00	0.01	0.03	0.07	0.11	0.16	0.20	0.25	0.30	0.35	0.40	0.46	0.50	0.55	0.59
1967	0.00	0.01	0.04	0.07	0.12	0.16	0.21	0.26	0.31	0.35	0.40	0.45	0.50	0.54	0.58
1968	0.00	0.01	0.04	0.08	0.12	0.17	0.22	0.27	0.31	0.36	0.41	0.45	0.50	0.54	0.59
1969	0.00	0.01	0.04	0.08	0.13	0.17	0.22	0.27	0.31	0.35	0.40	0.44	0.49	0.53	0.58
1970	0.00	0.02	0.04	0.08	0.13	0.18	0.22	0.27	0.31	0.35	0.39	0.44	0.48	0.53	0.57
1971	0.00	0.02	0.04	0.08	0.13	0.18	0.22	0.26	0.30	0.34	0.38	0.43	0.47	0.51	0.56
1972	0.00	0.02	0.04	0.08	0.13	0.18	0.22	0.26	0.29	0.33	0.37	0.42	0.46	0.50	0.54
1973	0.00	0.02	0.05	0.09	0.13	0.17	0.21	0.25	0.28	0.32	0.36	0.40	0.44	0.48	0.52
1974	0.00	0.02	0.05	0.08	0.13	0.17	0.21	0.24	0.28	0.32	0.36	0.40	0.44	0.48	0.52
1975	0.00	0.02	0.05	0.08	0.12	0.16	0.20	0.24	0.28	0.31	0.35	0.39	0.43	0.47	
1976	0.00	0.02	0.05	0.08	0.12	0.17	0.21	0.24	0.28	0.32	0.35	0.39	0.43		
1977	0.00	0.02	0.05	0.08	0.13	0.17	0.21	0.24	0.28	0.31	0.35	0.39			
1978	0.00	0.02	0.04	0.08	0.13	0.17	0.21	0.24	0.28	0.31	0.35				
1979	0.00	0.02	0.05	0.09	0.13	0.17	0.21	0.25	0.28	0.32					
1980	0.00	0.02	0.05	0.09	0.13	0.17	0.21	0.25	0.28						
1981	0.01	0.02	0.05	0.09	0.13	0.17	0.21	0.25							
1982	0.01	0.02	0.05	0.09	0.13	0.17	0.21								
1983	0.00	0.02	0.04	0.08	0.12	0.17									
1984	0.00	0.02	0.04	0.08	0.12										
1985	0.00	0.02	0.04	0.08											
1986	0.00	0.02	0.04												
1987	0.00	0.01													
1988	0.00														

1 See sections 2.9 and 3.9.
2 Includes births at ages 45 and over, achieved up to the end of 2003 by women born in 1958 and earlier years.

30	31	32	33	34	35	36	37	38	39	40	41	42	43	44	45²	Year of birth of woman/female birth cohort
0.70	0.72	0.74	0.75	0.76	0.77	0.77	0.78	0.78	0.79	0.79	0.79	0.79	0.79	0.79	0.79	1920
0.74	0.76	0.78	0.79	0.80	0.81	0.82	0.82	0.82	0.83	0.83	0.83	0.83	0.83	0.83	0.83	1925
0.74	0.77	0.78	0.80	0.81	0.82	0.82	0.83	0.83	0.83	0.83	0.84	0.84	0.84	0.84	0.84	1926
0.75	0.77	0.79	0.80	0.81	0.82	0.83	0.83	0.83	0.84	0.84	0.84	0.84	0.84	0.84	0.84	1927
0.76	0.78	0.80	0.81	0.82	0.83	0.84	0.84	0.84	0.85	0.85	0.85	0.85	0.85	0.85	0.85	1928
0.76	0.78	0.80	0.81	0.82	0.83	0.84	0.84	0.84	0.85	0.85	0.85	0.85	0.85	0.85	0.85	1929
0.78	0.80	0.82	0.83	0.84	0.85	0.85	0.86	0.86	0.86	0.86	0.86	0.87	0.87	0.87	0.87	1930
0.78	0.80	0.82	0.83	0.84	0.84	0.85	0.85	0.86	0.86	0.86	0.86	0.86	0.86	0.86	0.86	1931
0.78	0.80	0.82	0.83	0.84	0.84	0.85	0.85	0.86	0.86	0.86	0.86	0.86	0.86	0.86	0.86	1932
0.80	0.82	0.83	0.84	0.85	0.86	0.86	0.86	0.87	0.87	0.87	0.87	0.87	0.87	0.87	0.87	1933
0.82	0.84	0.85	0.86	0.87	0.87	0.88	0.88	0.88	0.88	0.89	0.89	0.89	0.89	0.89	0.89	1934
0.82	0.84	0.85	0.86	0.87	0.87	0.87	0.88	0.88	0.88	0.88	0.88	0.88	0.88	0.88	0.88	1935
0.82	0.84	0.85	0.86	0.86	0.87	0.87	0.88	0.88	0.88	0.88	0.88	0.88	0.88	0.88	0.88	1936
0.82	0.84	0.85	0.85	0.86	0.87	0.87	0.87	0.88	0.88	0.88	0.88	0.88	0.88	0.88	0.88	1937
0.83	0.84	0.85	0.86	0.87	0.87	0.88	0.88	0.88	0.88	0.88	0.88	0.89	0.89	0.89	0.89	1938
0.83	0.84	0.85	0.86	0.87	0.87	0.88	0.88	0.88	0.88	0.88	0.88	0.88	0.88	0.88	0.88	1939
0.83	0.85	0.86	0.87	0.87	0.88	0.88	0.88	0.88	0.89	0.89	0.89	0.89	0.89	0.89	0.89	1940
0.84	0.85	0.86	0.87	0.88	0.88	0.89	0.89	0.89	0.89	0.89	0.89	0.89	0.89	0.89	0.89	1941
0.84	0.85	0.86	0.87	0.87	0.88	0.88	0.88	0.89	0.89	0.89	0.89	0.89	0.89	0.89	0.89	1942
0.83	0.85	0.86	0.86	0.87	0.87	0.88	0.88	0.88	0.88	0.89	0.89	0.89	0.89	0.89	0.89	1943
0.84	0.86	0.87	0.87	0.88	0.89	0.89	0.89	0.90	0.90	0.90	0.90	0.90	0.90	0.90	0.90	1944
0.84	0.86	0.87	0.88	0.89	0.89	0.90	0.90	0.90	0.90	0.90	0.90	0.91	0.91	0.91	0.91	1945
0.84	0.86	0.87	0.88	0.89	0.90	0.90	0.90	0.91	0.91	0.91	0.91	0.91	0.91	0.91	0.91	1946
0.81	0.83	0.84	0.85	0.86	0.86	0.87	0.87	0.87	0.87	0.88	0.88	0.88	0.88	0.88	0.88	1947
0.81	0.83	0.85	0.86	0.86	0.87	0.88	0.88	0.88	0.88	0.89	0.89	0.89	0.89	0.89	0.89	1948
0.79	0.81	0.83	0.84	0.85	0.85	0.86	0.86	0.87	0.87	0.87	0.87	0.87	0.87	0.87	0.87	1949
0.78	0.80	0.82	0.83	0.84	0.84	0.85	0.85	0.86	0.86	0.86	0.86	0.86	0.86	0.86	0.86	1950
0.77	0.79	0.80	0.81	0.82	0.83	0.84	0.84	0.85	0.85	0.85	0.85	0.85	0.85	0.85	0.86	1951
0.76	0.78	0.80	0.81	0.82	0.83	0.84	0.84	0.85	0.85	0.85	0.85	0.85	0.85	0.85	0.85	1952
0.75	0.77	0.79	0.80	0.82	0.82	0.83	0.84	0.84	0.85	0.85	0.85	0.85	0.85	0.85	0.85	1953
0.73	0.76	0.77	0.79	0.80	0.81	0.82	0.83	0.83	0.83	0.84	0.84	0.84	0.84	0.84	0.84	1954
0.73	0.75	0.77	0.79	0.80	0.81	0.82	0.82	0.83	0.83	0.84	0.84	0.84	0.84	0.85	0.85	1955
0.71	0.74	0.76	0.78	0.79	0.80	0.81	0.82	0.82	0.83	0.83	0.83	0.84	0.84	0.84	0.84	1956
0.70	0.73	0.75	0.77	0.79	0.80	0.81	0.81	0.82	0.82	0.83	0.83	0.83	0.83	0.83	0.83	1957
0.69	0.72	0.74	0.76	0.77	0.79	0.79	0.80	0.81	0.81	0.82	0.82	0.82	0.82	0.82		1958
0.68	0.71	0.73	0.75	0.77	0.78	0.79	0.79	0.80	0.80	0.81	0.81	0.81	0.82			1959
0.67	0.70	0.73	0.74	0.76	0.77	0.78	0.79	0.80	0.80	0.81	0.81	0.81	0.81			1960
0.66	0.69	0.72	0.74	0.75	0.76	0.77	0.78	0.79	0.79	0.80	0.80	0.80				1961
0.66	0.69	0.71	0.73	0.75	0.76	0.77	0.78	0.79	0.79	0.80	0.80					1962
0.65	0.68	0.71	0.73	0.74	0.76	0.77	0.78	0.78	0.79	0.80						1963
0.64	0.67	0.69	0.72	0.74	0.75	0.76	0.77	0.78	0.78							1964
0.63	0.67	0.69	0.72	0.74	0.75	0.76	0.77	0.78								1965
0.63	0.66	0.69	0.72	0.74	0.75	0.76	0.77									1966
0.62	0.65	0.68	0.71	0.73	0.74	0.76										1967
0.63	0.66	0.69	0.72	0.74	0.75											1968
0.62	0.65	0.68	0.71	0.73												1969
0.61	0.64	0.68	0.70													1970
0.60	0.63	0.67														1971
0.58	0.62															1972
0.57																1973
																1974
																1975
																1976
																1977
																1978
																1979
																1980
																1981
																1982
																1983
																1984
																1985
																1986
																1987
																1988

Table 10.4 Components of average family size[1]: occurrence within/outside marriage, birth order, mother's year of birth, and age of mother at birth, 1920-1984   **England and Wales**

Mother's year of birth	All live births	Outside marriage	Within marriage					
			Birth order					
			All	First	Second	Third	Fourth	Fifth and later
	All ages of mothers at birth							
1920	**2.00**	0.13	1.87	0.76	0.55	0.28	0.14	0.15
1925	**2.12**	0.12	2.00	0.79	0.58	0.31	0.15	0.17
1930	**2.35**	0.12	2.23	0.84	0.66	0.36	0.18	0.19
1935	**2.42**	0.13	2.29	0.85	0.71	0.38	0.18	0.16
1940	**2.36**	0.16	2.21	0.85	0.73	0.37	0.16	0.11
1941	**2.34**	0.16	2.18	0.85	0.73	0.36	0.15	0.10
1942	**2.29**	0.16	2.12	0.84	0.72	0.34	0.13	0.08
1943	**2.24**	0.16	2.08	0.84	0.72	0.33	0.12	0.07
1944	**2.21**	0.16	2.04	0.84	0.72	0.31	0.11	0.06
1945	**2.19**	0.17	2.02	0.85	0.72	0.30	0.10	0.05
1946	**2.19**	0.18	2.01	0.85	0.72	0.29	0.10	0.05
1947	**2.08**	0.17	1.91	0.82	0.69	0.27	0.09	0.04
1948	**2.11**	0.18	1.93	0.82	0.71	0.27	0.09	0.04
1949	**2.08**	0.19	1.89	0.81	0.69	0.27	0.08	0.04
1950	**2.07**	0.20	1.87	0.80	0.69	0.26	0.08	0.04
1951	**2.04**	0.20	1.84	0.78	0.67	0.26	0.08	0.04
1952	**2.05**	0.21	1.84	0.78	0.67	0.26	0.08	0.05
1953	**2.05**	0.22	1.82	0.78	0.66	0.26	0.08	0.05
1954	**2.02**	0.23	1.79	0.76	0.65	0.25	0.08	0.05
1955	**2.02**	0.25	1.77	0.75	0.65	0.25	0.08	0.05
1956	**2.02**	0.27	1.75	0.74	0.64	0.25	0.08	0.05
1957	**2.01**	0.29	1.73	0.72	0.63	0.25	0.08	0.05
1958	**1.99**	0.31	1.68	0.71	0.61	0.24	0.08	0.04
	Under 20							
1920	**0.08**	0.01	0.06	0.06	0.01	0.00		
1925	**0.08**	0.02	0.06	0.05	0.01	0.00		
1930	**0.10**	0.02	0.09	0.08	0.01	0.00		
1935	**0.11**	0.02	0.09	0.08	0.01	0.00		
1940	**0.16**	0.03	0.13	0.11	0.02	0.00		
1941	**0.17**	0.03	0.14	0.12	0.02	0.00		
1942	**0.18**	0.03	0.14	0.12	0.02	0.00		
1943	**0.19**	0.03	0.15	0.12	0.02	0.00		
1944	**0.20**	0.04	0.16	0.13	0.03	0.00		
1945	**0.22**	0.04	0.17	0.14	0.03	0.00		
1946	**0.22**	0.05	0.17	0.14	0.03	0.00		
1947	**0.22**	0.05	0.17	0.13	0.03	0.00		
1948	**0.22**	0.05	0.17	0.13	0.03	0.00		
1949	**0.23**	0.06	0.17	0.13	0.03	0.00		
1950	**0.23**	0.06	0.17	0.14	0.03	0.00		
1951	**0.24**	0.06	0.18	0.14	0.03	0.00		
1952	**0.25**	0.06	0.18	0.15	0.03	0.00		
1953	**0.24**	0.06	0.18	0.14	0.03	0.00		
1954	**0.23**	0.06	0.17	0.14	0.03	0.00		
1955	**0.22**	0.06	0.16	0.13	0.03	0.00		
1956	**0.20**	0.06	0.14	0.11	0.03	0.00		
1957	**0.19**	0.06	0.13	0.10	0.02	0.00		
1958	**0.17**	0.06	0.11	0.09	0.02	0.00		
1959	**0.16**	0.06	0.10	0.08	0.02	0.00		
1960	**0.15**	0.06	0.09	0.08	0.02	0.00		
1961	**0.16**	0.06	0.09	0.08	0.02	0.00		
1962	**0.15**	0.07	0.09	0.07	0.02	0.00		
1963	**0.14**	0.07	0.07	0.06	0.01	0.00		
1964	**0.13**	0.07	0.06	0.05	0.01	0.00		
1965	**0.13**	0.08	0.06	0.04	0.01	0.00		
1966	**0.13**	0.08	0.05	0.04	0.01	0.00		
1967	**0.14**	0.09	0.05	0.04	0.01	0.00		
1968	**0.15**	0.10	0.04	0.03	0.01	0.00		
1969	**0.15**	0.11	0.04	0.03	0.01	0.00		
1970	**0.15**	0.12	0.04	0.03	0.01	0.00		

Note: Average family sizes are obtained by summing rates for each single year of age.
1 See sections 2.9, 2.11 and 3.9.

Table 10.4 - *continued*

Mother's year of birth	All live births	Outside marriage	Within marriage					
			Birth order					
			All	First	Second	Third	Fourth	Fifth and later
			Under 20 - *continued*					
1971	**0.15**	0.12	0.03	0.02	0.01	0.00		
1972	**0.15**	0.13	0.03	0.02	0.01	0.00		
1973	**0.15**	0.13	0.03	0.02	0.01	0.00		
1974	**0.15**	0.13	0.02	0.02	0.01	0.00		
1975	**0.15**	0.13	0.02	0.02	0.00	0.00		
1976	**0.15**	0.13	0.02	0.02	0.00	0.00		
1977	**0.15**	0.13	0.02	0.02	0.00	0.00		
1978	**0.15**	0.13	0.02	0.01	0.00	0.00		
1979	**0.15**	0.14	0.02	0.01	0.00	0.00		
1980	**0.15**	0.14	0.02	0.01	0.00	0.00		
1981	**0.15**	0.14	0.02	0.01	0.00	0.00		
1982	**0.15**	0.13	0.02	0.01	0.00	0.00		
1983	**0.14**	0.13	0.01	0.01	0.00	0.00		
1984	**0.14**	0.13	0.01	0.01	0.00	0.00		
	20-24							
1920	**0.49**	0.04	0.45	0.31	0.11	0.03	0.01	0.00
1925	**0.63**	0.05	0.58	0.38	0.15	0.04	0.01	0.00
1930	**0.66**	0.03	0.63	0.38	0.18	0.05	0.01	0.00
1935	**0.76**	0.04	0.72	0.43	0.21	0.06	0.02	0.00
1940	**0.89**	0.06	0.83	0.44	0.27	0.09	0.03	0.01
1941	**0.91**	0.06	0.84	0.44	0.28	0.09	0.03	0.01
1942	**0.90**	0.07	0.83	0.43	0.28	0.09	0.02	0.01
1943	**0.87**	0.06	0.81	0.42	0.27	0.09	0.02	0.01
1944	**0.86**	0.07	0.79	0.41	0.27	0.08	0.02	0.01
1945	**0.84**	0.07	0.77	0.40	0.27	0.08	0.02	0.01
1946	**0.82**	0.07	0.76	0.39	0.26	0.08	0.02	0.00
1947	**0.77**	0.06	0.71	0.37	0.24	0.07	0.02	0.00
1948	**0.76**	0.06	0.70	0.37	0.25	0.07	0.02	0.00
1949	**0.72**	0.06	0.66	0.35	0.23	0.06	0.01	0.00
1950	**0.69**	0.06	0.63	0.33	0.23	0.06	0.01	0.00
1951	**0.65**	0.05	0.59	0.31	0.21	0.05	0.01	0.00
1952	**0.62**	0.05	0.56	0.29	0.21	0.05	0.01	0.00
1953	**0.58**	0.05	0.53	0.28	0.20	0.05	0.01	0.00
1954	**0.56**	0.05	0.51	0.26	0.19	0.04	0.01	0.00
1955	**0.55**	0.05	0.50	0.26	0.18	0.04	0.01	0.00
1956	**0.55**	0.06	0.50	0.26	0.18	0.04	0.01	0.00
1957	**0.55**	0.06	0.49	0.26	0.17	0.04	0.01	0.00
1958	**0.54**	0.07	0.47	0.25	0.17	0.04	0.01	0.00
1959	**0.53**	0.08	0.45	0.25	0.16	0.04	0.01	0.00
1960	**0.52**	0.09	0.44	0.23	0.15	0.04	0.01	0.00
1961	**0.50**	0.09	0.41	0.22	0.14	0.04	0.01	0.00
1962	**0.48**	0.10	0.38	0.20	0.13	0.03	0.01	0.00
1963	**0.47**	0.11	0.35	0.19	0.12	0.03	0.01	0.00
1964	**0.46**	0.13	0.33	0.18	0.11	0.03	0.01	0.00
1965	**0.46**	0.14	0.31	0.17	0.11	0.03	0.01	0.00
1966	**0.46**	0.16	0.30	0.16	0.10	0.03	0.01	0.00
1967	**0.45**	0.18	0.28	0.15	0.09	0.03	0.01	0.00
1968	**0.45**	0.19	0.26	0.14	0.09	0.03	0.01	0.00
1969	**0.44**	0.20	0.24	0.13	0.08	0.02	0.00	0.00
1970	**0.42**	0.20	0.22	0.12	0.08	0.02	0.00	0.00
1971	**0.40**	0.21	0.20	0.10	0.07	0.02	0.00	0.00
1972	**0.39**	0.21	0.18	0.10	0.06	0.02	0.00	0.00
1973	**0.38**	0.21	0.17	0.09	0.06	0.02	0.00	0.00
1974	**0.38**	0.22	0.16	0.08	0.06	0.02	0.00	0.00
1975	**0.37**	0.22	0.15	0.08	0.05	0.02	0.00	0.00
1976	**0.37**	0.22	0.14	0.08	0.05	0.01	0.00	0.00
1977	**0.36**	0.22	0.14	0.07	0.05	0.01	0.00	0.00
1978	**0.35**	0.22	0.13	0.07	0.04	0.01	0.00	0.00
1979	**0.35**	0.22	0.13	0.07	0.04	0.01	0.00	0.00

Table 10.4 - *continued*

Mother's year of birth	All live births	Outside marriage	Within marriage					
			Birth order					
			All	First	Second	Third	Fourth	Fifth and later
25-29								
1920	**0.70**	0.04	0.66	0.28	0.24	0.10	0.04	0.02
1925	**0.67**	0.02	0.65	0.23	0.24	0.11	0.04	0.03
1930	**0.78**	0.02	0.76	0.26	0.27	0.13	0.06	0.04
1935	**0.90**	0.04	0.87	0.26	0.32	0.17	0.07	0.05
1940	**0.85**	0.04	0.80	0.22	0.31	0.16	0.07	0.04
1941	**0.81**	0.04	0.77	0.22	0.30	0.16	0.06	0.04
1942	**0.79**	0.04	0.75	0.22	0.29	0.15	0.06	0.03
1943	**0.76**	0.04	0.73	0.22	0.29	0.14	0.05	0.03
1944	**0.73**	0.03	0.70	0.22	0.29	0.13	0.04	0.02
1945	**0.71**	0.03	0.68	0.22	0.28	0.12	0.04	0.02
1946	**0.69**	0.03	0.66	0.23	0.28	0.11	0.03	0.01
1947	**0.64**	0.03	0.62	0.22	0.26	0.09	0.03	0.01
1948	**0.64**	0.03	0.61	0.22	0.26	0.09	0.03	0.01
1949	**0.62**	0.03	0.59	0.22	0.25	0.08	0.02	0.01
1950	**0.62**	0.03	0.59	0.22	0.25	0.09	0.02	0.01
1951	**0.63**	0.03	0.59	0.23	0.25	0.09	0.02	0.01
1952	**0.64**	0.04	0.60	0.23	0.25	0.09	0.03	0.01
1953	**0.65**	0.04	0.61	0.24	0.25	0.09	0.03	0.01
1954	**0.65**	0.04	0.61	0.23	0.24	0.09	0.03	0.01
1955	**0.65**	0.05	0.60	0.23	0.24	0.09	0.03	0.01
1956	**0.64**	0.05	0.59	0.22	0.24	0.09	0.03	0.01
1957	**0.64**	0.06	0.58	0.22	0.23	0.09	0.03	0.01
1958	**0.64**	0.07	0.57	0.22	0.22	0.09	0.03	0.01
1959	**0.63**	0.07	0.56	0.22	0.22	0.08	0.02	0.01
1960	**0.63**	0.08	0.55	0.22	0.21	0.08	0.03	0.01
1961	**0.62**	0.09	0.53	0.21	0.20	0.08	0.02	0.01
1962	**0.61**	0.10	0.51	0.21	0.19	0.07	0.02	0.01
1963	**0.61**	0.11	0.50	0.21	0.19	0.07	0.02	0.01
1964	**0.60**	0.12	0.48	0.20	0.18	0.06	0.02	0.01
1965	**0.59**	0.13	0.46	0.20	0.17	0.06	0.02	0.01
1966	**0.58**	0.14	0.44	0.19	0.16	0.06	0.02	0.01
1967	**0.56**	0.15	0.42	0.18	0.15	0.06	0.02	0.01
1968	**0.55**	0.15	0.40	0.18	0.15	0.05	0.02	0.01
1969	**0.54**	0.16	0.38	0.17	0.14	0.05	0.02	0.01
1970	**0.52**	0.16	0.36	0.17	0.13	0.05	0.02	0.01
1971	**0.51**	0.17	0.34	0.16	0.12	0.04	0.01	0.01
1972	**0.49**	0.16	0.32	0.15	0.11	0.04	0.01	0.01
1973	**0.47**	0.16	0.31	0.15	0.11	0.04	0.01	0.01
1974	**0.47**	0.17	0.30	0.15	0.10	0.04	0.01	0.00
30-34								
1920	**0.43**	0.02	0.41	0.09	0.14	0.09	0.05	0.05
1925	**0.45**	0.02	0.43	0.09	0.14	0.10	0.05	0.06
1930	**0.53**	0.02	0.51	0.09	0.15	0.12	0.07	0.08
1935	**0.47**	0.02	0.44	0.07	0.13	0.11	0.06	0.06
1940	**0.35**	0.02	0.33	0.06	0.11	0.09	0.05	0.04
1941	**0.33**	0.02	0.31	0.06	0.10	0.08	0.04	0.03
1942	**0.30**	0.02	0.29	0.06	0.10	0.07	0.03	0.02
1943	**0.29**	0.02	0.27	0.06	0.10	0.07	0.03	0.02
1944	**0.28**	0.02	0.27	0.06	0.11	0.06	0.03	0.02
1945	**0.29**	0.02	0.28	0.06	0.11	0.07	0.03	0.01
1946	**0.31**	0.02	0.29	0.07	0.12	0.07	0.03	0.01
1947	**0.31**	0.02	0.30	0.07	0.12	0.07	0.02	0.01
1948	**0.34**	0.02	0.32	0.08	0.13	0.08	0.03	0.01
1949	**0.35**	0.02	0.33	0.08	0.13	0.08	0.03	0.01
1950	**0.36**	0.02	0.34	0.08	0.14	0.08	0.03	0.01
1951	**0.36**	0.03	0.33	0.08	0.13	0.08	0.03	0.02
1952	**0.37**	0.03	0.34	0.08	0.14	0.08	0.03	0.02
1953	**0.38**	0.03	0.35	0.09	0.14	0.08	0.03	0.02
1954	**0.39**	0.04	0.36	0.09	0.14	0.08	0.03	0.02
1955	**0.40**	0.04	0.36	0.10	0.14	0.08	0.03	0.02

Note: Average family sizes are obtained by summing rates for each single year of age.

Table 10.4 - *continued*

Mother's year of birth	All live births	Outside marriage	Within marriage					
			Birth order					
			All	First	Second	Third	Fourth	Fifth and later
30-34 - *continued*								
1956	**0.41**	0.05	0.36	0.10	0.14	0.08	0.03	0.02
1957	**0.42**	0.06	0.36	0.10	0.14	0.08	0.03	0.02
1958	**0.43**	0.06	0.37	0.10	0.14	0.08	0.03	0.02
1959	**0.43**	0.07	0.37	0.10	0.14	0.08	0.03	0.02
1960	**0.44**	0.07	0.37	0.11	0.15	0.07	0.03	0.01
1961	**0.44**	0.08	0.36	0.11	0.14	0.07	0.02	0.01
1962	**0.44**	0.08	0.36	0.11	0.14	0.07	0.02	0.01
1963	**0.45**	0.09	0.36	0.11	0.15	0.07	0.02	0.01
1964	**0.45**	0.10	0.36	0.11	0.15	0.07	0.02	0.01
1965	**0.46**	0.10	0.36	0.12	0.15	0.06	0.02	0.01
1966	**0.45**	0.10	0.35	0.12	0.14	0.06	0.02	0.01
1967	**0.45**	0.11	0.34	0.12	0.14	0.06	0.02	0.01
1968	**0.45**	0.11	0.34	0.12	0.14	0.05	0.02	0.01
1969	**0.46**	0.11	0.34	0.12	0.14	0.05	0.02	0.01
35 and over								
1920	**0.30**	0.02	0.28	0.03	0.06	0.06	0.05	0.08
1925	**0.30**	0.02	0.28	0.03	0.06	0.06	0.05	0.08
1930	**0.26**	0.02	0.25	0.03	0.05	0.06	0.04	0.07
1935	**0.17**	0.01	0.16	0.02	0.03	0.04	0.03	0.04
1940	**0.12**	0.01	0.11	0.02	0.03	0.03	0.02	0.02
1941	**0.12**	0.01	0.11	0.02	0.03	0.03	0.02	0.02
1942	**0.12**	0.01	0.11	0.02	0.03	0.03	0.02	0.02
1943	**0.12**	0.01	0.11	0.02	0.03	0.03	0.02	0.02
1944	**0.13**	0.01	0.12	0.02	0.03	0.03	0.02	0.02
1945	**0.13**	0.01	0.12	0.02	0.03	0.03	0.02	0.02
1946	**0.14**	0.02	0.12	0.02	0.03	0.03	0.02	0.02
1947	**0.14**	0.02	0.12	0.02	0.03	0.03	0.02	0.01
1948	**0.14**	0.02	0.13	0.02	0.04	0.03	0.02	0.02
1949	**0.15**	0.02	0.13	0.02	0.04	0.03	0.02	0.02
1950	**0.16**	0.02	0.14	0.02	0.04	0.04	0.02	0.02
1951	**0.17**	0.03	0.14	0.03	0.04	0.04	0.02	0.02
1952	**0.18**	0.03	0.15	0.03	0.05	0.04	0.02	0.02
1953	**0.19**	0.03	0.15	0.03	0.05	0.04	0.02	0.02
1954	**0.19**	0.04	0.16	0.03	0.05	0.04	0.02	0.02
1955	**0.20**	0.04	0.16	0.03	0.05	0.04	0.02	0.02
1956	**0.21**	0.04	0.16	0.04	0.05	0.04	0.02	0.02
1957	**0.21**	0.05	0.16	0.04	0.06	0.04	0.02	0.02
1958	**0.22**	0.05	0.17	0.04	0.06	0.04	0.02	0.02

Table 10.5 Estimated distribution of women of child-bearing age by number of live-born children (percentages)[1]: year of birth and age, 1920-1983   England and Wales

Year of birth of woman	Age of woman (completed years)	Number of live-born children[2]				
		0 (Childless women)	1	2	3	4 or more
1920	20	89	9	2	0	0
1925		88	11	2	0	0
1930		84	13	3	0	0
1935		83	13	3	0	0
1940		78	16	4	1	0
1945		74	18	6	1	0
1950		74	18	7	1	0
1955		77	16	6	1	0
1960		82	13	4	1	0
1965		85	11	4	1	0
1970		82	13	3	1	0
1975		84	13	3	1	0
1980		83	14	3	0	0
1983		83	13	3	0	0
1920	25	55	28	12	4	1
1925		46	32	16	5	2
1930		44	30	17	6	2
1935		39	30	21	7	3
1940		34	27	25	9	5
1945		34	27	27	9	4
1950		41	24	25	7	2
1955		48	22	22	6	2
1960		55	20	18	6	2
1965		60	19	15	5	1
1970		61	20	14	5	2
1975		65	17	13	4	1
1978		65	18	12	4	1
1920	30	30	28	25	11	6
1925		26	29	27	12	7
1930		22	26	30	14	9
1935		18	21	32	17	12
1940		17	17	35	19	12
1945		16	19	41	16	8
1950		22	19	40	14	6
1955		27	19	35	13	5
1960		33	18	31	12	5
1965		37	20	28	11	5
1970		39	22	24	10	4
1973		43	21	23	9	4
1920	35	23	23	28	15	11
1925		19	24	29	16	12
1930		15	20	31	18	16
1935		13	16	32	21	18
1940		12	14	36	22	16
1945		11	15	43	20	11
1950		16	14	43	18	9
1955		19	15	40	18	9
1960		23	14	37	18	9
1965		25	16	36	16	8
1968		25	18	34	15	8
1920	40	21	22	27	16	14
1925		17	22	28	17	16
1930		14	18	30	19	19
1935		12	15	32	21	20
1940		11	13	36	22	18
1945		10	14	43	21	12
1950		14	13	44	19	10
1955		16	13	41	19	10
1960		19	12	39	19	10
1963		20	13	39	18	9
1920	45[3]	21	21	27	16	15
1925		17	22	28	17	16
1930		13	18	30	19	20
1935		12	15	32	21	20
1940		11	13	36	22	18
1945		9	14	43	21	12
1950		14	13	44	20	11
1955		15	13	41	20	10
1958		18	13	39	20	11

Note: Figures may not add exactly due to rounding - see section 2.15.
1   See sections 2.9 and 3.9.
2   Estimates including births both within and outside marriage - see section 2.9.
3   Includes births at ages over 45.

**Table 11.1   Live births within marriage: estimated distribution
by socio-economic classification of father as defined by occupation,
by number of previous live-born children and age of mother, 2001-2003**

England and Wales

thousands

Year	Number of previous live-born children					Year	Number of previous live-born children				
	Total	0	1	2	3 or more		Total	0	1	2	3 or more
Socio-economic classification 1.1						**Socio-economic classification 2**					
All ages of mother at birth						**All ages of mother at birth**					
2001	**34.1**	14.8	13.6	4.4	1.3	2001	**82.5**	36.4	31.0	11.1	4.0
2002	**35.1**	15.0	14.1	4.5	1.5	2002	**81.1**	35.9	30.9	10.0	4.2
2003	**35.7**	15.3	14.3	4.7	1.5	2003	**86.1**	38.6	32.4	10.8	4.3
Under 20						**Under 20**					
2001	**0.1**	0.0	0.0	0.0	0.0	2001	**0.5**	0.4	0.1	0.0	0.0
2002	**0.0**	0.0	0.0	0.0	0.0	2002	**0.5**	0.5	0.1	0.0	0.0
2003	**0.0**	0.0	0.0	0.0	0.0	2003	**0.4**	0.3	0.0	0.0	0.0
20-24						**20-24**					
2001	**1.1**	0.8	0.3	0.1	0.0	2001	**6.3**	4.0	1.9	0.4	0.0
2002	**1.1**	0.7	0.4	0.0	0.0	2002	**6.3**	3.9	1.9	0.4	0.1
2003	**1.2**	0.7	0.4	0.0	0.0	2003	**6.6**	4.1	1.9	0.5	0.1
25-29						**25-29**					
2001	**7.7**	4.5	2.4	0.7	0.1	2001	**22.7**	12.7	7.1	2.0	0.7
2002	**7.1**	4.2	2.2	0.5	0.1	2002	**22.1**	12.6	7.1	1.9	0.6
2003	**7.1**	4.2	2.2	0.5	0.2	2003	**22.1**	12.4	7.2	1.7	0.7
30 and over						**30 and over**					
2001	**25.1**	9.4	10.8	3.7	1.2	2001	**52.9**	19.1	22.0	8.6	3.3
2002	**26.9**	10.0	11.5	4.0	1.4	2002	**52.2**	19.0	21.8	7.7	3.6
2003	**27.5**	10.4	11.6	4.2	1.3	2003	**57.1**	21.8	23.2	8.6	3.6
Socio-economic classification 1.2						**Socio-economic classification 3**					
All ages of mother at birth						**All ages of mother at birth**					
2001	**47.3**	21.1	17.7	6.0	2.4	2001	**21.1**	9.5	8.0	2.4	1.0
2002	**47.8**	22.4	17.4	5.8	2.1	2002	**21.4**	10.4	7.5	2.5	1.0
2003	**49.5**	22.9	18.3	6.3	2.0	2003	**22.4**	10.7	8.5	2.4	0.8
Under 20						**Under 20**					
2001	**0.1**	0.1	0.0	0.0	0.0	2001	**0.4**	0.4	0.1	0.0	0.0
2002	**0.1**	0.1	0.0	0.0	0.0	2002	**0.3**	0.3	0.0	0.0	0.0
2003	**0.1**	0.1	0.0	0.0	0.0	2003	**0.2**	0.2	0.0	0.0	0.0
20-24						**20-24**					
2001	**2.2**	1.6	0.5	0.1	0.0	2001	**2.7**	1.7	0.8	0.2	0.0
2002	**2.0**	1.4	0.5	0.1	0.0	2002	**2.7**	1.8	0.7	0.1	0.1
2003	**2.2**	1.5	0.6	0.1	0.0	2003	**3.0**	1.8	1.0	0.2	0.0
25-29						**25-29**					
2001	**10.6**	6.4	3.3	0.6	0.3	2001	**6.2**	3.1	2.3	0.5	0.2
2002	**10.4**	6.5	3.0	0.7	0.2	2002	**6.5**	3.6	2.0	0.7	0.2
2003	**10.9**	7.0	3.1	0.6	0.2	2003	**6.8**	3.8	2.2	0.6	0.2
30 and over						**30 and over**					
2001	**34.2**	12.9	13.9	5.3	2.1	2001	**11.8**	4.3	4.9	1.8	0.8
2002	**35.2**	14.3	13.8	5.1	1.9	2002	**11.9**	4.7	4.8	1.7	0.7
2003	**36.4**	14.3	14.7	5.6	1.8	2003	**12.3**	4.9	5.2	1.6	0.7

Note: For a description of socio-economic classifications and the sample used in calculation - see section 3.10.

Table 11.1 - *continued*   *thousands*

Year	Number of previous live-born children					Year	Number of previous live-born children				
	Total	0	1	2	3 or more		Total	0	1	2	3 or more
Socio-economic classification 4						**Socio-economic classification 6**					
All ages of mother at birth						**All ages of mother at birth**					
2001	**41.9**	14.1	15.5	7.2	5.1	2001	**33.6**	12.6	11.8	5.6	3.6
2002	**43.8**	14.7	15.9	8.2	5.1	2002	**34.0**	13.0	11.6	5.4	4.0
2003	**44.8**	15.5	16.1	8.2	5.0	2003	**31.8**	11.7	11.1	5.1	3.9
Under 20						**Under 20**					
2001	**0.6**	0.5	0.1	0.0	0.0	2001	**0.9**	0.8	0.1	0.0	0.0
2002	**0.5**	0.4	0.1	0.0	0.0	2002	**1.0**	0.8	0.2	0.0	0.0
2003	**0.5**	0.4	0.1	0.0	0.0	2003	**0.9**	0.7	0.2	0.0	0.0
20-24						**20-24**					
2001	**4.7**	2.4	1.7	0.6	0.1	2001	**7.8**	3.7	2.9	0.9	0.3
2002	**4.8**	2.3	1.8	0.6	0.1	2002	**7.8**	4.1	2.7	0.9	0.2
2003	**5.1**	2.6	1.7	0.6	0.2	2003	**6.8**	3.3	2.6	0.7	0.2
25-29						**25-29**					
2001	**12.0**	4.7	4.7	1.8	0.9	2001	**11.2**	4.2	4.0	2.1	0.9
2002	**11.6**	4.5	4.1	2.0	0.9	2002	**10.7**	4.1	3.7	1.8	1.1
2003	**12.3**	4.8	4.4	2.2	0.9	2003	**9.9**	3.8	3.3	1.7	1.1
30 and over						**30 and over**					
2001	**24.6**	6.6	9.1	4.8	4.1	2001	**13.8**	4.0	4.8	2.7	2.4
2002	**26.9**	7.5	9.8	5.6	4.0	2002	**14.5**	4.0	5.1	2.6	2.7
2003	**26.9**	7.7	9.9	5.4	3.9	2003	**14.0**	3.9	4.9	2.6	2.6
Socio-economic classification 5						**Socio-economic classification 7**					
All ages of mother at birth						**All ages of mother at birth**					
2001	**45.2**	17.4	17.8	6.5	3.5	2001	**38.4**	13.4	13.2	6.5	5.3
2002	**43.3**	17.2	16.9	5.9	3.3	2002	**33.2**	11.2	11.9	5.6	4.5
2003	**41.9**	16.8	15.9	5.8	3.3	2003	**34.9**	13.0	11.5	5.9	4.4
Under 20						**Under 20**					
2001	**0.5**	0.4	0.1	0.0	0.0	2001	**1.0**	0.8	0.2	0.0	0.0
2002	**0.4**	0.4	0.0	0.0	0.0	2002	**1.0**	0.8	0.2	0.0	0.0
2003	**0.5**	0.4	0.1	0.0	0.0	2003	**0.9**	0.8	0.2	0.0	0.0
20-24						**20-24**					
2001	**5.6**	3.0	2.0	0.4	0.1	2001	**7.3**	3.4	2.7	0.9	0.2
2002	**5.6**	2.9	2.0	0.6	0.1	2002	**7.2**	3.5	2.7	0.9	0.2
2003	**5.6**	2.8	2.1	0.6	0.1	2003	**6.8**	3.3	2.5	0.7	0.3
25-29						**25-29**					
2001	**15.6**	6.8	6.0	1.9	0.9	2001	**13.1**	4.9	4.6	2.4	1.3
2002	**14.0**	6.2	5.5	1.5	0.7	2002	**10.5**	3.6	3.8	1.9	1.2
2003	**12.8**	5.9	4.5	1.8	0.6	2003	**11.4**	4.4	3.8	2.2	1.1
30 and over						**30 and over**					
2001	**23.5**	7.2	9.6	4.2	2.5	2001	**17.1**	4.4	5.7	3.3	3.7
2002	**23.3**	7.7	9.4	3.8	2.4	2002	**14.5**	3.3	5.3	2.8	3.1
2003	**22.9**	7.7	9.2	3.4	2.7	2003	**15.7**	4.5	5.0	3.0	3.1

Note: For a description of socio-economic classifications and the sample used in calculation - see section 3.10.

Table 11.1 - *continued*

Year	Number of previous live-born children					Year	Number of previous live-born children				
	Total	0	1	2	3 or more		**Total**	0	1	2	3 or more
	Socio-economic classification 8						**All socio-economic classifications (including 'not classified')**				
	All ages of mother at birth						**All ages of mother at birth**				
2001	**0.1**	0.0	0.0	0.0	0.0	2001	**356.5**	143.9	132.2	52.1	28.3
2002	**0.1**	0.0	0.0	0.0	0.0	2002	**354.1**	145.2	130.3	50.3	28.2
2003	**0.0**	0.0	0.0	0.0	0.0	2003	**364.2**	151.0	132.9	52.0	28.4
	Under 20						**Under 20**				
2001	**0.0**	0.0	0.0	0.0	0.0	2001	**4.6**	3.8	0.8	0.1	0.0
2002	**0.0**	0.0	0.0	0.0	0.0	2002	**4.6**	3.8	0.7	0.1	0.0
2003	**0.0**	0.0	0.0	0.0	0.0	2003	**4.3**	3.5	0.8	0.1	0.0
	20-24						**20-24**				
2001	**0.0**	0.0	0.0	0.0	0.0	2001	**40.7**	22.2	13.7	3.9	0.9
2002	**0.0**	0.0	0.0	0.0	0.0	2002	**40.7**	22.4	13.5	3.9	0.9
2003	**0.0**	0.0	0.0	0.0	0.0	2003	**40.9**	22.2	13.9	3.8	1.0
	25-29						**25-29**				
2001	**0.0**	0.0	0.0	0.0	0.0	2001	**103.1**	48.8	35.6	12.8	5.9
2002	**0.0**	0.0	0.0	0.0	0.0	2002	**97.6**	47.1	33.0	11.8	5.6
2003	**0.0**	0.0	0.0	0.0	0.0	2003	**98.7**	48.4	32.5	12.1	5.7
	30 and over						**30 and over**				
2001	**0.0**	0.0	0.0	0.0	0.0	2001	**208.0**	69.1	82.1	35.4	21.5
2002	**0.0**	0.0	0.0	0.0	0.0	2002	**211.2**	71.9	83.2	34.5	21.6
2003	**0.0**	0.0	0.0	0.0	0.0	2003	**220.3**	76.9	85.7	36.0	21.8
	Not classified										
	All ages of mother at birth										
2001	**12.4**	4.5	3.5	2.3	2.1						
2002	**14.4**	5.5	4.0	2.3	2.5						
2003	**17.1**	6.4	4.8	2.7	3.1						
	Under 20										
2001	**0.6**	0.5	0.1	0.0	0.0						
2002	**0.6**	0.5	0.1	0.0	0.0						
2003	**0.7**	0.6	0.1	0.0	0.0						
	20-24										
2001	**3.1**	1.6	0.9	0.4	0.1						
2002	**3.3**	1.9	0.9	0.4	0.1						
2003	**3.6**	2.0	1.2	0.3	0.1						
	25-29										
2001	**3.9**	1.3	1.2	0.8	0.6						
2002	**4.6**	1.7	1.5	0.9	0.6						
2003	**5.2**	1.9	1.6	0.9	0.8						
	30 and over										
2001	**5.0**	1.1	1.3	1.1	1.4						
2002	**5.8**	1.3	1.6	1.1	1.8						
2003	**7.4**	1.9	1.9	1.4	2.2						

Note: For a description of socio-economic classifications and the sample used in calculation - see section 3.10.

Table 11.2 Pre-maritally conceived first live births to married women (numbers and percentages): estimated distribution by socio-economic classification of father as defined by occupation, and age of mother, 2001-2003

England and Wales

Year	All socio-economic classifications (including 'not classified')	Socio-economic classification of father								
		1.1	1.2	2	3	4	5	6	7	8
	Number (thousands)									
	All ages of mother at birth									
2001	**13.6**	1.1	1.4	3.4	1.1	1.2	1.8	1.2	1.6	0.0
2002	**13.9**	1.0	1.5	3.0	1.1	1.7	1.8	1.6	1.4	0.0
2003	**14.5**	1.2	1.5	2.9	1.0	1.7	1.9	1.4	1.9	0.0
	Under 20									
2001	**1.1**	0.0	0.0	0.1	0.1	0.1	0.1	0.2	0.2	0.0
2002	**1.1**	0.0	0.0	0.1	0.1	0.1	0.1	0.2	0.3	0.0
2003	**1.0**	0.0	0.0	0.1	0.1	0.1	0.1	0.2	0.2	0.0
	20-24									
2001	**3.8**	0.2	0.2	0.9	0.4	0.3	0.6	0.4	0.5	0.0
2002	**3.6**	0.1	0.1	0.7	0.4	0.4	0.5	0.6	0.6	0.0
2003	**3.8**	0.1	0.1	0.6	0.4	0.4	0.7	0.4	0.7	0.0
	25-29									
2001	**4.1**	0.3	0.5	1.0	0.3	0.4	0.5	0.4	0.5	0.0
2002	**4.1**	0.4	0.5	0.9	0.2	0.5	0.7	0.4	0.3	0.0
2003	**4.1**	0.3	0.5	1.0	0.2	0.4	0.5	0.4	0.5	0.0
	30 and over									
2000	**4.6**	0.5	0.7	1.4	0.2	0.4	0.6	0.3	0.4	0.0
2001	**5.1**	0.5	0.8	1.3	0.4	0.7	0.5	0.4	0.3	0.0
2003	**5.6**	0.7	0.9	1.2	0.3	0.7	0.6	0.4	0.5	0.0
	As a percentage of all first live births within marriage									
	All ages of mother at birth									
2001	**9.4**	7.2	6.8	9.3	11.4	8.8	10.4	9.9	11.9	0.0
2002	**9.6**	6.9	6.6	8.4	10.8	11.3	10.2	12.4	12.7	0.0
2003	**9.6**	7.6	6.7	7.5	9.0	10.6	11.2	11.7	14.8	-
	Under 20									
2001	**28.7**	67.4	33.7	30.8	39.3	22.5	30.0	30.4	27.0	-
2002	**29.3**	48.9	30.1	28.8	38.5	24.5	22.0	28.8	33.0	0.0
2003	**29.8**	0.0	39.5	21.1	42.2	33.6	26.5	23.4	31.8	-
	20-24									
2001	**16.9**	25.9	11.6	21.7	23.3	14.0	19.0	11.1	15.8	0.0
2002	**16.1**	20.3	7.6	18.0	21.2	16.8	17.0	13.6	16.2	0.0
2003	**17.2**	20.5	9.7	15.7	19.4	15.9	22.9	11.9	19.9	-
	25-29									
2001	**8.4**	6.5	7.6	7.7	9.7	8.0	7.9	8.5	10.4	0.0
2002	**8.6**	8.4	8.0	7.0	5.6	10.0	11.1	10.3	8.8	-
2003	**8.4**	7.7	6.4	8.0	5.8	7.6	8.8	11.4	11.1	-
	30 and over									
2001	**6.7**	5.6	5.5	7.2	5.7	6.5	8.2	6.5	8.1	-
2002	**7.1**	5.2	5.7	6.9	9.1	9.7	6.3	9.9	8.4	-
2003	**7.2**	6.7	6.3	5.5	6.6	9.7	8.1	9.4	11.8	-

Note: For a description of socio-economic classifications and the sample used in calculation - see section 3.10.

Table 11.3 Median birth intervals: socio-economic classification of father as defined by occupation (for interval to first birth), and mother's marriage order, 2001-2003

United Kingdom[2]
England and Wales

Year	Median intervals in months[1]										First to second birth	Second to third birth	Third to fourth birth
	Marriage to first birth												
	All socio-economic classifications (including 'not classified') 1.1	1.2	2	3	4	5	6	7	8				
	Women married once only (England and Wales)										All women (United Kingdom)[1,2]		
2001	**27**	32	30	28	25	25	26	25	23	12	36	41	38
2002	**26**	32	31	28	25	24	27	23	21	12	37	40	37
2003	**26**	30	30	27	25	24	25	23	24	-	37	42	38
	Remarried women (England and Wales)												
2001	**18**	19	19	18	19	20	17	16	22	-			
2002	**17**	18	16	19	21	17	16	14	22	-			
2003	**18**	19	19	20	20	16	17	18	15	-			

Note: For a description of socio-economic classifications and the sample used in calculation - see section 3.10.
1 See section 3.11.
2 Incorrectly labelled 'Great Britain' in the 2002 volume.

Table 11.4 Mean age of women at live births within marriage: socio-economic classification of father as defined by occupation, and birth order[1], 2001-2003

England and Wales

Year	All socio-economic classifications (including 'not classified')	Socio-economic classification of father								
		1.1	1.2	2	3	4	5	6	7	8
	All live births within marriage									
2001	**30.9**	32.6	32.5	31.7	30.5	31.1	30.4	28.9	29.4	26.5
2002	**31.0**	32.8	32.7	31.6	30.5	31.2	30.4	29.0	29.2	25.1
2003	**31.2**	32.9	32.6	31.8	30.5	31.2	30.5	29.2	29.3	33.2
	First live births within marriage									
2001	**29.7**	31.5	31.1	30.4	29.2	29.7	29.3	27.5	27.8	26.3
2002	**29.7**	31.7	31.5	30.3	29.2	29.9	29.3	27.3	27.1	20.5
2003	**29.9**	31.7	31.3	30.6	29.4	29.9	29.4	27.6	27.9	-
	Second live births within marriage									
2001	**31.2**	33.1	33.0	32.2	31.0	30.9	30.4	28.8	29.2	26.8
2002	**31.4**	33.2	33.2	32.2	31.2	31.2	30.6	29.2	29.3	25.5
2003	**31.5**	33.3	33.2	32.3	31.1	31.3	30.6	29.1	29.2	26.5
	Third live births within marriage									
2001	**32.4**	34.2	34.8	33.6	32.7	32.2	32.1	30.2	30.4	-
2002	**32.3**	34.6	34.5	33.1	32.6	32.3	31.7	30.2	30.4	29.0
2003	**32.5**	34.7	34.9	33.4	31.9	32.2	31.3	30.5	30.3	34.5

Notes: For a description of socio-economic classifications and the sample used in calculation - see section 3.10.
The mean ages shown in this table are not standardised and therefore take no account of the structure of the population by age, marital status or parity.
1 See section 2.9.

Table 11.5   **Jointly registered live births outside marriage: socio-economic**   **England and Wales**
classification of father as defined by occupation, and age of mother, 2001-2003   *thousands*

Year	Age of mother	All socio-economic classifications (including 'not classified')	Socio-economic classification of father								
			1.1	1.2	2	3	4	5	6	7	8
2001	**All ages**	**194.3**	**7.3**	**9.4**	**29.6**	**9.3**	**25.6**	**33.5**	**28.3**	**40.7**	**0.0**
2002		**198.9**	**7.6**	**9.3**	**30.0**	**9.6**	**28.1**	**33.8**	**28.4**	**40.2**	**0.1**
2003		**212.3**	**8.3**	**10.3**	**32.2**	**11.0**	**30.4**	**34.6**	**28.8**	**42.6**	**0.1**
2001	Under 20	**27.9**	0.2	0.3	1.9	1.1	2.1	4.6	5.6	8.3	0.0
2002		**27.7**	0.2	0.4	1.9	1.0	2.2	3.9	5.6	8.4	0.0
2003		**28.7**	0.2	0.4	1.9	1.2	2.3	4.3	5.4	8.4	0.0
2001	20-24	**54.5**	0.9	1.4	6.8	2.5	6.3	10.2	9.3	13.7	0.0
2002		**56.6**	1.1	1.4	6.3	2.8	6.8	11.3	9.8	13.5	0.0
2003		**61.3**	1.0	1.4	7.3	3.2	7.7	10.6	10.0	15.5	0.0
2001	25-29	**48.2**	1.9	2.4	7.7	2.6	6.8	8.8	6.9	9.4	0.0
2002		**47.7**	1.9	2.3	7.9	2.5	7.3	8.5	6.3	9.0	0.0
2003		**49.7**	2.1	2.3	8.1	2.7	7.6	8.9	6.4	9.1	0.0
2001	30 and over	**63.7**	4.3	5.3	13.2	3.2	10.4	10.0	6.5	9.2	0.0
2002		**66.9**	4.4	5.2	13.8	3.3	11.9	10.2	6.7	9.4	0.0
2003		**72.7**	5.1	6.3	14.9	3.9	12.8	10.8	6.9	9.6	0.0

Note: For a description of socio-economic classifications and the sample used in calculation - see section 3.10.

Appendix Table 1   Estimated resident population[1]: sex and age, 1993-2003

England and Wales

thousands

| | Age | Year | | | | | | | | | | |
		1993	1994	1995	1996	1997	1998	1999	2000	2001	2002	2003
Persons	**All ages**	**50,985.9**	**51,116.2**	**51,272.0**	**51,410.4**	**51,559.6**	**51,720.1**	**51,933.5**	**52,140.2**	**52,360.0**	**52,570.2**	**52,793.7**
Males	All ages	24,792.6	24,853.0	24,946.3	25,029.8	25,113.2	25,200.7	25,323.4	25,438.0	25,574.3	25,702.4	25,840.6
Females	All ages	26,193.2	26,263.2	26,325.6	26,380.7	26,446.5	26,519.4	26,610.0	26,702.1	26,785.7	26,867.8	26,953.1
	15–44	**10,748.9**	**10,720.9**	**10,711.9**	**10,717.6**	**10,727.0**	**10,737.2**	**10,766.6**	**10,820.8**	**10,869.5**	**10,905.2**	**10,944.1**
	15–19	1,458.9	1,454.1	1,473.8	1,501.4	1,534.3	1,560.7	1,566.5	1,562.7	1,578.4	1,609.3	1,649.6
	20–24	1,842.6	1,774.3	1,712.0	1,633.0	1,560.6	1,516.6	1,516.9	1,540.0	1,577.7	1,604.4	1,636.8
	25–29	2,062.9	2,041.3	2,004.9	1,979.7	1,945.2	1,902.3	1,850.9	1,809.9	1,744.7	1,674.0	1,627.5
	30–34	1,956.2	2,008.0	2,051.0	2,075.8	2,087.9	2,080.7	2,069.2	2,049.3	2,033.4	2,010.3	1,974.5
	35–39	1,723.7	1,760.0	1,802.5	1,853.9	1,903.4	1,951.9	2,004.0	2,053.7	2,082.0	2,102.5	2,101.0
	40–44	1,704.6	1,683.3	1,667.7	1,673.8	1,695.6	1,725.0	1,759.1	1,805.2	1,853.3	1,904.8	1,954.6
	45–49	1,768.9	1,802.5	1,821.8	1,825.4	1,744.4	1,693.2	1,673.9	1,664.6	1,670.1	1689.7	1,719.9

Note: Figures may not add exactly due to rounding - see section 2.15.
1 See section 2.1.

Appendix Table 2   Estimated resident female population[1]: age and marital status, 1992–2003

England and Wales

thousands

| | Age | Year | | | | | | | | | | | |
		1992	1993	1994	1995	1996	1997	1998	1999	2000	2001	2002	2003
Married	**15-44**	**5,584.9**	**5,429.3**	**5,309.2**	**5,178.1**	**5,070.5**	**4,962.6**	**4,859.5**	**4,772.1**	**4,696.7**	**4,611.1**	**4,490.8**	**4,372.4**
	15-19	27.2	23.7	21.6	20.8	20.7	20.3	20.3	19.8	17.6	16.2	13.0	11.7
	20-24	439.2	392.0	345.0	300.0	259.6	224.9	201.1	187.8	179.6	178.0	166.1	160.8
	25-29	1,114.8	1,068.1	1,022.9	962.7	906.3	844.3	782.8	725.1	677.2	625.4	567.4	523.5
	30-34	1,334.8	1,334.3	1,336.4	1,330.4	1,315.8	1,293.1	1,259.4	1,222.8	1,182.2	1,142.4	1,094.5	1,042.6
	35-39	1,289.7	1,290.9	1,297.1	1,305.5	1,319.9	1,331.5	1,342.0	1,354.1	1,362.1	1,356.3	1,342.4	1,314.6
	40-44	1,379.3	1,320.4	1,287.0	1,258.7	1,248.1	1,248.4	1,253.9	1,262.5	1,278.0	1,292.9	1,307.4	1,319.1
	45-49	1,350.0	1,399.0	1,412.8	1,413.5	1,402.1	1,323.0	1,268.4	1,239.3	1,218.7	1,208.1	1,206.0	1,211.8
Single, widowed and divorced	**15-44**	**5,264.6**	**5,319.6**	**5,411.7**	**5,533.9**	**5,647.1**	**5,764.4**	**5,877.7**	**5,994.6**	**6,124.1**	**6,258.4**	**6,414.4**	**6,571.7**
	15-19	1,483.0	1,435.2	1,432.5	1,453.1	1,480.7	1,514.0	1,540.4	1,546.8	1,545.1	1,562.3	1,596.3	1,637.9
	20-24	1,457.4	1,450.6	1,429.3	1,412.0	1,373.3	1,335.7	1,315.6	1,329.1	1,360.5	1,399.7	1,438.3	1,476.0
	25-29	967.3	994.8	1,019.0	1,042.2	1,073.5	1,100.8	1,119.4	1,125.8	1,132.7	1,119.3	1,106.6	1,104.0
	30-34	574.7	622.0	671.6	720.6	760.0	794.7	821.3	846.4	867.1	891.0	915.8	931,8
	35-39	405.8	432.9	463.0	497.0	534.0	571.9	609.9	649.8	691.6	725.7	760.0	786.5
	40-44	376.4	384.2	396.4	409.0	425.7	447.3	471.1	496.7	527.3	560.5	597.4	635.5
	45-49	343.7	369.9	389.8	408.3	423.3	421.4	424.8	434.6	446.0	462.0	483.7	508.1

Note: Figures may not add exactly due to rounding - see section 2.15.
1 See section 2.1.

Appendix Table 3   **Estimated standard errors for numbers in analyses by socio-economic classification**

Estimated number[1] (thousands)	Standard error (thousands)	Standard error as a percentage of estimated number
0.5	0.07	14.0
1.0	0.09	9.0
5.0	0.21	4.2
10.0	0.30	3.0
20.0	0.42	2.1
30.0	0.50	1.7
40.0	0.58	1.4
50.0	0.64	1.3
60.0	0.69	1.2
70.0	0.74	1.1
80.0	0.78	1.0
90.0	0.82	0.9
100.0	0.86	0.9
150.0	0.99	0.7
200.0	1.07	0.5
250.0	1.10	0.4
300.0	1.10	0.4

1 Numbers relate to those estimated in Section 11 (socio-economic classification) of this volume from a ten per cent sample.

Appendix Table 4   **Estimated standard errors for percentages in analyses by socio-economic classification**

Estimated number[1] on which percentage is based (thousands)	Percentages					
	5 or 95	10 or 90	20 or 80	30 or 70	40 or 60	50
0.5	2.9	4.0	5.4	6.1	6.6	6.7
1.0	2.1	2.8	3.8	4.3	4.6	4.7
5.0	0.9	1.3	1.7	1.9	2.1	2.1
10.0	0.7	0.9	1.2	1.4	1.5	1.5
20.0	0.5	0.6	0.8	1.0	1.0	1.1
30.0	0.4	0.5	0.7	0.8	0.8	0.9
40.0	0.3	0.4	0.6	0.7	0.7	0.8
50.0	0.3	0.4	0.5	0.6	0.7	0.7
60.0	0.3	0.4	0.5	0.6	0.6	0.6
70.0	0.2	0.3	0.5	0.5	0.6	0.6
80.0	0.2	0.3	0.4	0.5	0.5	0.5
90.0	0.2	0.3	0.4	0.5	0.5	0.5
100.0	0.2	0.3	0.4	0.4	0.5	0.5
150.0	0.2	0.2	0.3	0.4	0.4	0.4
200.0	0.1	0.2	0.3	0.3	0.3	0.3
250.0	0.1	0.2	0.2	0.3	0.3	0.3
300.0	0.1	0.2	0.2	0.3	0.3	0.3

1 Numbers relate to those estimated in Section 11 (socio-economic classification) of this volume from a ten per cent sample.

Annex A   Draft entry form used currently for registering live births (Form 309(Rev))

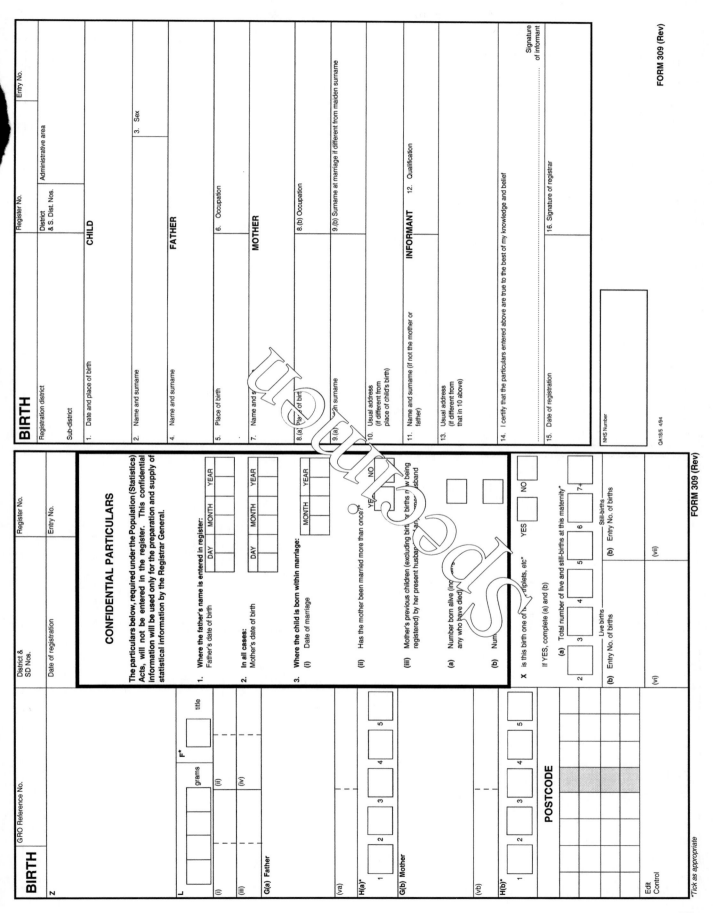

Annex B

Draft entry form used currently for registering stillbirths (Form 308(Rev))

STILL-BIRTH

Register No. | **Entry No.**

Registration district

District & S. Dist. Nos. | Administrative area

Sub-district

CHILD

1.(a) Date and place of birth

3. Sex

1.(b) Name and surname

2. Cause of death and nature of evidence that child was still-born

a

b

c

d

e

Certified

FATHER

4. Name and surname

5. Place of birth

6. Occupation

MOTHER

7. Name and surname

8.(a) Place of birth

8.(b) Occupation

9.(a) Maiden surname

9.(b) Surname at marriage if different from maiden surname

10. Usual address (if different from place of child's birth)

INFORMANT

11. Name and surname (if not the mother or father)

12. Qualification

13. Usual address (if different from that in 10 above)

14. I certify that the particulars entered above are true to the best of my knowledge and belief

.... Signature of informant

15. Date of registration

16. Signature of registrar

OA1/8/3 5/98

FORM 308

STILL-BIRTH

District & SD Nos. | **Register No.**

Date of registration | Entry No.

(i) (ii) grams (iii) weeks

K   2   4   6

(iv) ME

N Post Mortem Held?*   Yes   No

U

Y*   Before Labour a   During Labour b   Not Known c

CONFIDENTIAL PARTICULARS

The particulars below, required under the Population (Statistics) Acts, will not be entered in the register. This confidential information will be used only for the preparation and supply of statistical information by the Registrar General.

1. Where the father's name is entered in register:
   Father's date of birth   DAY MONTH YEAR

2. In all cases:
   Mother's date of birth   DAY MONTH YEAR

3. Where the child is born within marriage:
   (I) Date of marriage   MONTH YEAR
   (II) Has the mother been married more than once?*   Yes   No
   (III) Mother's previous ... birth or births now ... present ...
   (a) Number ... live ...
   (b) Number ... still-born

X is this birth one of twins, triplets, etc*   YES   NO

If YES, complete (a) and (b)

(a) Total number of live and still-births at this maternity*
   2 3 4 5 6 7+

Live births
(b) Entry No. of births

Still-births
(b) Entry No. of births

(x)   (xi)

(v) a
b
c
d
e

M

(vi)

(vii)

(viii)

(ixa)

(ixb)

G(a) Father

H(a)* 1 2 3 4 5

G(b) Mother

H(b)* 1 2 3 4 5

POSTCODE

Edit Control

*Tick as appropriate

FORM 308